Donald C. Hylton
Understanding Plastics Testing

Donald C. Hylton

Understanding
Plastics Testing

HANSER

Hanser Publishers, Munich • Hanser Gardner Publications, Cincinnati

The Author:
Donald C. Hylton
223 James P. Brawley Drive, Atlanta, GA 30314, USA

Distributed in the USA and in Canada by
Hanser Gardner Publications, Inc.
6915 Valley Avenue, Cincinnati, Ohio 45244-3029, USA
Fax: (513) 527-8801
Phone: (513) 527-8977 or 1-800-950-8977
Internet: http://www.hansergardner.com

Distributed in all other countries by
Carl Hanser Verlag
Postfach 86 04 20, 81631 München, Germany
Fax: +49 (89) 98 48 09
Internet: http://www.hanser.de

The use of general descriptive names, trademarks, etc., in this publication, even if the former are not especially identified, is not to be taken as a sign that such names, as understood by the Trade Marks and Merchandise Marks Act, may accordingly be used freely by anyone.

While the advice and information in this book are believed to be true and accurate at the date of going to press, neither the authors nor the editors nor the publisher can accept any legal responsibility for any errors or omissions that may be made. The publisher makes no warranty, express or implied, with respect to the material contained herein.

Library of Congress Cataloging-in-Publication Data
Hylton, Donald C.
Understanding plastics testing / Donald C. Hylton.-- 1st ed.
p. cm.
Includes index.
ISBN 1-56990-366-2 (pbk.)
1. Plastics-Testing. I. Title.
TA455.P5H95 2004
620.1'923--dc22
2004014755

Bibliografische Information Der Deutschen Bibliothek
Die Deutsche Bibliothek verzeichnet diese Publikation in der DeutschenNationalbibliografie;
detaillierte bibliografische Daten sind im Internetüber <http://dnb.ddb.de> abrufbar.
ISBN 3-446-22246-4

© Carl Hanser Verlag, Munich 2004
Production Management: Oswald Immel
Typeset by Manuela Treindl, Laaber, Germany
Coverconcept: Marc Müller-Bremer, Rebranding, München, Germany
Coverdesign: MCP • Susanne Kraus GbR, Holzkirchen, Germany
Printed and bound by Druckhaus "Thomas Müntzer", Bad Langensalza, Germany

Preface

The main objective of this primer on plastics testing is to present the subject in a manner that is understandable by the unlearned yet unchallenged by the learned. The plastics industry with its continued growth, demands of increased quality awareness, and versatility of applications, has forced testing of plastics to become essential to the continued success of the industry. It is important to know how to conduct a test but it is even more important to know what information the test yields and does not yield. This book presents a brief overview of materials, their properties, and attributes that contribute to results obtained from testing.

The primer can be separated into three basic segments. It begins by presenting an overview of material technology. The emphasis is on understanding plastics properties and behavior and how they influence testing. The second segment presents some overviews of popular tests, their background, and uses. It is not possible in the scope of this work to present a summary of all tests. The tests covered here represent the tests that are most frequently included on specifications and data sheets. Some tests such as creep and thermal analysis are included because of the very important information they can provide.

The final segment presents some concepts on quality and quality assurance. No testing laboratory is complete unless it places quality as its number one priority. Quality principles and practices demand thorough knowledge of testing procedures and understandings of the information generated.

Donald C. Hylton
June 2004

Acknowledgements

I owe any insight I may have to my teacher and mentor, Dr. Costel D. Denson. He took the time to personally teach me about rheology. With that he instilled in me a unique and simplistic way of thinking and approaching such a very difficult and complex subject as rheology. He taught me and explained to me what appear to be extraordinary complex mathematical principles. It was this teaching that established a base for my understanding of polymers and their behavior.

Throughout my career, it has been necessary to develop new test methods for new understandings of polymer properties or to select materials for specific applications. From this I developed a love for material testing and became a constant presence in laboratories.

Bill McConnell, the well known and respected plastic consultant challenged me to undertake this effort a few years ago and I have been struggling to meet that challenge since. The effort has led to a series of seminars and to this endeavor.

Contents

1 The Science of Testing

1.1 Why Test?

In today's information age, customers are more aware of the products they are using and the standards they must meet to safely and reliably be consumed. In the past, testing of materials were afterthoughts and a nuisance. However, technological and scientific advances have made testing a must, because with knowledge come responsibility and liability. Manufacturers are required to meet a number of specifications, standards, health and safety demands, and to provide information about their products.

Plastics use in products has grown substantially in recent years. The driving forces are cost reduction, automation, and high yield. This has forced the plastic industry into placing extra emphasis on testing and developing new procedures. More importantly, it has forced the plastics engineers, designers, scientists, and technicians to develop an increased understanding of plastics, their capabilities, limitations, and properties.

1.2 Meeting the Standard – American Society for Testing and Materials International (ASTM)

The efforts of organizations on standards, suppliers, and primarily the American Society of Testing and Materials (ASTM) have resulted in thousands of methods and procedures for testing plastics. Fortunately, ASTM has emerged as the premier organization that organizes, standardizes, and documents practically all known tests for plastics. As a result, most of the published plastics testing and practices can be found in published volumes of ASTM standards. ASTM volumes include standard test methods, specifications, practices, guides, classifications, and terminology in areas covering plastics and plastic-based materials such as paints, textiles, consumer products, computerized systems, electronics, and many others areas involved in everyday life.

ASTM International is a non-profit organization founded in 1898. It provides a global forum for the development and publication of standards and test methods. Its membership is comprised of producers, users, consumers, and representatives of government and academia. ASTM International provides standards that are accepted and used in research and development, product testing, quality systems, and commercial transactions around the globe.

Within ASTM, the primary responsibility for plastics lies with a committee designated for this purpose. This committee, called D-20 on Plastics, is responsible for more than 500 standard test methods, recommended practices, and guides. One of the key components of D-20 is the continuous review and updating of existing documents and the authoring of new protocols that are necessary.

New standards development begins when committee members identify or any interested party identifies a need. A task group prepares a draft of the standard, which is reviewed and submitted to the parent subcommittee for a balloting process. After approval by the subcommittee, the document is submitted concurrently to the main committee (D-20) and the Society. All members are provided an opportunity to vote. Negative votes must be submitted in writing, fully explaining the voter's objection. A negative vote anywhere in the process must be documented and resolved fully before the document can be submitted to the next level or become a standard.

Subcommittees are charged with reviewing existing documents every five years. The review process examines a document for its current relevancy, wording, applicability, precision, and accuracy. The reviewed document is then subjected to the exact same balloting process mentioned in the preceding paragraph.

Some other organizations that have major influences on plastics standards and testing are:

- National Institute of Standards and Technology (NIST)

- International Organization for Standardization (ISO)

- Underwriters Laboratory (UL)

- Society of Plastics Industry (SPI)

- Department of Defense

- American National Standards Institute (ANSI)

- American Society for Quality Control (ASQC)

- American Society of Electroplated Plastics

- American Society of Safety Engineers

- EUROLAB

- National Association of Plastics Fabricators

- Federal Drug Administration (FDA)

- National Science Foundation (NSF)

- United States Department of Agriculture (USDA)

2   Understanding Polymers and Their Behavior

2.1   Basic Polymer Science

Before one can begin to understand plastics testing one must first have a basic concept of the characteristics of the material tested and why they behave in the manner observed. Plastics are very complex materials in that their behavior is governed by chemistry, architecture, molecular weight, distribution of molecular weights, physical state, configuration, and use mechanisms. In this section an attempt will be made to demystify and explain these characteristics in an elementary and picturesque manner.

Plastics commonly refer to a manmade material called a polymer. To be specific, the word *plastic* is a descriptive term referring to a substance that is capable of being shaped, reshaped, formed, molded, and drawn without losing its basic functionality.

The word *polymer* is a combination of the words "*poly*" meaning many and the word "*mer*" meaning unit. In the world of plastics, a polymer is a chemical substance produced by combining many (poly) small molecules (mers) together to make one very large molecule.

The process of producing polymers is called *polymerization*. In polymerization, *monomers* (single units) are introduced into a reaction vessel under ideal temperature and pressure conditions. Usually, a *catalyst* is used to facilitate the reaction. Temperature, pressure, catalyst, monomer, and the amount of monomer control what is called reaction conditions. These reaction conditions produce a polymer with specific properties and uses called *attributes*.

These attributes are chemical structure, molecular weight, molecular weight distribution, and architecture (*morphology*). Polymer attributes are a result of how the polymer is manufactured. They dictate how polymers can be processed, how they can be used, and what properties they will demonstrate when tested. Some specific attributes will be described in this chapter and applied to understanding behavior in subsequent chapters.

2.2   Polymer Chemistry

Monomers are small molecules that contain carbon atoms. Some compounds containing carbon atoms under certain conditions are capable of combining with themselves to produce a maxi-molecular reproduction of the smaller starting unit. Carbon atoms have a valence of four. This means that four other atoms or molecules can attach themselves to a carbon atom under the right conditions. Some examples are hydrogen, oxygen, nitrogen, carbon, benzene, hydrocarbons, and others. Thus the primary building block for polymers is the carbon atom. The base monomer normally contains a minimum of two attached carbon atoms with hydrogen filling one to two of the valences of each carbon atom and a distinguishing compound, let's call "R". The two open valences remaining will allow the carbon atom to attach to another monomer (Fig. 2.1).

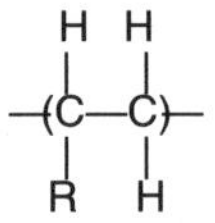

$$
\begin{array}{cc}
\text{H} & \text{H} \\
| & | \\
-(\text{C}\!-\!\text{C})- \\
| & | \\
\text{R} & \text{H}
\end{array}
$$

Figure 2.1 Basic building block for common polymers

The R groups are primarily responsible for the unique properties of a polymer. R side groups containing $-CH_2-$ pendants or chains are generally flexible, and may be more crystalline. Polymers containing aromatic side groups, such as a benzene ring, are more stiff and hard but can be brittle. Table 2.1 is a list of some common pendants and their contribution to physical properties.

Table 2.1 Properties Resulting from Side Chain Pendant Groups

R – Pendant	Polymper Example	Properties
H	Polyethylene HDPE	Flexible, soft, chemically resistant, opaque, crystalline, high density, short chain branching
CH_3	Polypropylene	Rigid, chemical resistant, opaque, highly crystalline
$-(CH_2)_n-$	Polyethylene LDPE	Flexible, soft, chemical resistant, low crystallinity, low density, long chain branching
C_6H_5	Polystyrene	Clear, strong, brittle, low chemical resistance, non-crystalline
Cl	Polyvinyl-Chloride	Clear, strong, low burning, non crystalline, poor heat stability

Table 2.2 Properties Resulting from Heterochain Groups in Polymer Backbones

R – Backbone group	Example	Properties
–N–	Polyamide, Nylon	Crystalline tough, smooth, poor water resistance, high mechanical strength resistant to wear and tear, excellent shape stability
–COO–	PET	Hard, stiff, strong, dimensionally stable, absorbs very little water, good gas barrier properties, good chemical resistance except to alkalis
–C$_6$H$_5$–	Polycarbonate	High strength, excellent impact, clear, durable, high temperature resistance. Excellent optical properties

In some instances, R groups can be substituted directly into the carbon-to-carbon chain rather than existing as a side group or pendant. These polymers are called heterochain polymers. Hetereochain polymers are more complex and possess unique properties. Table 2.2 lists some common examples.

As can be seen, chemical structure is also used to give the material its generic name. For example, the polymer from the ethylene monomer is polyethylene, the one from the propylene monomer is polypropylene, and so forth.

A common test for chemical structure is Fourier Transform Infrared Spectroscopy (FTIR). This is a complicated technique that requires highly trained scientists to interpret the data. Simplistically, it is a measurement of the wavelength at which infrared rays are absorbed or transmitted, resulting from energy changes with specific chemical bonds and linkages. For example, carbon-to-carbon bonds (C–C) will absorb at a unique wavelength. Other types of bonds (C=C, C–N, etc) will likewise have their unique wavelengths. The result will be a series of peaks called a spectrum on a chart recorder. By running a spectrum, one can get a fingerprint of a given polymer and identify an unknown material.

FTIR spectra are also used in some polymer processes to fine-tune radiant heating devices to a specific wavelength to achieve maximum efficiency when heating materials to be processed.

2.3   Molecular Weight and Molecular Weight Distribution

During commercial polymerization, it is extremely difficult to produce a polymer system in which all molecules are of the same molecular weight. The result is a system of molecules with varying molecular weights and chain lengths. This distribution of molecular weights is extremely important in determining polymer properties.

Molecular weight and molecular weight distribution are very important in determining a material's flow properties. They also affect physical and mechanical properties by influencing the final state of a solid material.

Most investigations are concerned with the weight average molecular weight, M_w, and the number average molecular weight, M_n. These are determined by counting and adding all the molecules of a given size. It is represented by using the following expressions

$$M_n = \frac{\sum N_i \cdot M_i}{\sum N_i} \tag{2.1}$$

and

$$M_w = \frac{\sum (N_i \cdot M_i) M_i}{\sum M_i \cdot N_i} \tag{2.2}$$

Where N_i is the number of molecules of molecular weight M_i, and $\sum$ represents adding them all together. It is generally concluded that M_n represents low molecular weight components while M_w is regarded as the average molecular weight of the system. Figure 2.2 illustrates a typical manner in which molecular weight data is reported. The variable M_Z in the figure represents high end molecular weights

Gel Permeation chromatography (GPC) is a standard method for determining molecular weight and molecular weight distribution. Polymers are dissolved into a diluted solution and injected into a tube containing a substance that filters and separates the different molecules according to their size. A count is taken and presented as the number of occurrences at a given place within the tube. It is graphically presented as distribution curve similar to Fig. 2.2.

Molecular weight distribution, MWD, is normally quantified by the ratio of the weight average molecular weight to the number average molecular weight M_w / M_n. When MWD approaches 1, the system is considered as having a narrow molecular weight

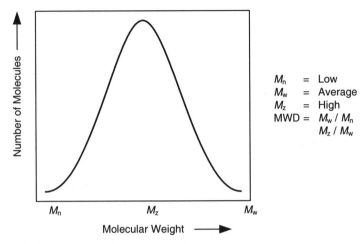

Figure 2.2 Typical molecular weight distribution data

distribution. In the unlikely event of MWD = 1 the system is *monodisperse,* that is, all molecules are of the same size. This does not normally occur in commercial processes, rather there will be a distribution of differing molecular sizes (polydisperse). This can be desirable as the different molecular weights impart unique and synergistic behaviors to the polymer. For example, low molecular weight improves processing and high molecular weight improves mechanical properties.

2.4   Polymer Architecture or Morphology

Polymers are long thin chains consisting of millions of monomer units. These chains are packed together in a random fashion and intermingled with another. Imagine a pot of spaghetti: when at rest, the spaghetti lengths are randomly oriented. When forces are applied such as stirring, the spaghetti strands tend to align themselves in the direction they are stirred. When heated, they become more flexible and easier to stir. When cooled, they become more rigid and adhere together. Stirring becomes more difficult as the strands resist aligning with the stirring direction. Polymers behave in much the same manner.

There are four major categories of plastics according to their molecular chain structure (Fig. 2.3).

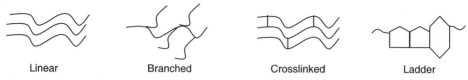

Linear   Branched   Crosslinked   Ladder

Figure 2.3 Basic polymer structures

1. Linear – Long random coil chain lengths where the side chain pendant groups, R, are single or short chain carbon atoms or aromatic groups.

2. Branched – The molecule consists of a main polymer chain while the side chains R are long chain polymer groups chemically attached to the main backbone.

3. Crosslinked – The molecules are formed by linear chains chemically connected by short chain groups forming a three-dimensional network.

4. Ladder – Aromatic ring groups tightly stacked and interconnected.

The most common types of commercial polymers are linear or branched polymers, on which most of the discussions in this work will be focused. Crosslinked polymers are used extensively when high strength and high temperature properties are required.

A solid section of plastic material consists of numerous long polymer chains stacked and interlocked with each other. In some cases, the polymer chains are randomly interlocked while in other cases they may be more orderly. The manner in which the chains align is a major contributor to behavior and performance. If the chains are aligned in a random disordered fashion the material is called **amorphous**. Amorphous materials do not have a finite melting point and usually are transparent.

If the polymer chains are stacked in an orderly fashion the material is considered to be **crystalline**. Crystalline materials as characterized by having a finite melting point. They are rigid and opaque. X-Ray crystallography is a standard technique for determining crystallinity. Thermal analysis (to be discussed later) also is an acceptable technique for determining crystallinity.

Another important polymer attribute is **tacticity** (see Fig. 2.4). Tacticity represents the arrangement of the side pendants along the polymer backbone. If the pendants alternate, the polymer is considered **syndiotactic**, see Fig. 2.4a. Recent catalyst technology permits the production of syndiotactic polypropylene, which is clear and generally non crystalline. If the pendants are hanging on the same side, the arrangement is called **isotactic**, see Fig. 2.4b. As expected, this arrangement is very orderly and will facilitate a crystalline structure. This is the most common commercial form of polypropylene,

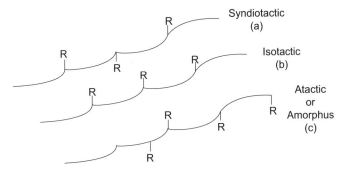

Figure 2.4 Illustration of polymer tacticity

where the methyl groups on the polypropylene molecule are all located on the same side of the chain. Another common form of structure is called **atactic**, see Fig. 2.4c. In the atactic arrangement the pendant locations are random. These types of polymers are non-crystalline and clear. Polystyrene is a good example of a polymer with an atactic arrangement of its benzene side groups.

Not only can monomers react with themselves, they can also react with other monomers forming a **copolymer**. A copolymer is a material containing dissimilar monomers in the backbone that are chemically bonded. They can combine in a random fashion forming a **random copolymer,** as shown in Fig. 2.5a. In some cases the monomers will react with themselves, forming a polymer chain and then react with each other, forming a **block copolymer** (see Fig. 2.5b). A third copolymer type occurs when a polymer chain bonds as a side pendant with a main polymer chain forming what is known as a **graft copolymer,** as shown in Fig. 2.5c.

In order to meet increasing demands for specific applications for polymers, polymer scientists have formulated numerous **blends** and **alloys** of the many commercial polymers known today. Blending and alloying can modify a material to render it useful in applications in which it would otherwise not be useful. One such example is impact modified polystyrene. Polystyrene is a strong rigid material; however, it is brittle and will break very easily. By adding a small amount of rubber to the material, an otherwise brittle material will become increasingly tougher while retaining its rigidity.

```
AABABBABBAAA          AAAABBBBAAAABBBB          AAAAAAAAAAAA
                                                B          B
                                                B          B
                                                B          B
                                                B          B
                                                B          B
```

  (a) Random   (b) Block   (c) Graft

Figure 2.5 Examples of copolymers

There are many types and variations of blends and alloys, in fact they are too numerous to name or illustrate them all in this work.

Through the years, polymer scientists have taken advantage of the many different ways in which polymers can be modified. Few commercial materials are as versatile, chemically as well as physically, as polymers.

2.5   Polymer Rheology

To gain a true understanding of plastics testing one must be introduced to and have a basic understanding of polymer rheology. Rheology is the science of studying how materials respond to applied forces, internally as well as externally. Practically all physical property testing involves some form of rheological phenomena. In essence, physical testing involves bending, pulling, twisting, pressing, heating, cooling, and others. In all cases, the forces are applied and the material's response is measured. Therefore, this work will devote considerable effort to seeking an understanding of the basic fundamentals of polymer rheology.

2.5.1   Deformation, Stress, and Strain

Polymer chains are flexible and will move when forces are applied. They will align in the direction of the force and become less random. As the force continues to be applied, the polymer molecules become less random and will begin to move or flow. When heat is applied this process becomes easier, requiring less force. We will refer to this chain motion as *deformation*.

There are three basic mechanisms of chain deformation (see Fig. 2.6). The first type is the bending and stretching of the bonds between the atoms within the polymer molecule, as shown in Fig. 2.6a. This type of deformation occurs first, it is very small and is *elastic* in nature, that is, the molecule will completely recover to its original position instantly. The second mode of deformation is the uncoiling or straightening of the coiled polymer chain (see Fig. 2.6b). This deformation is also elastic in nature but the recovery is not instantaneous, rather it occurs over a period of time as the straight molecules seek a more desired random coil position. The third mode is the slippage or flow of the molecules relative to one another (see Fig. 2.6c). This type of deformation is permanent

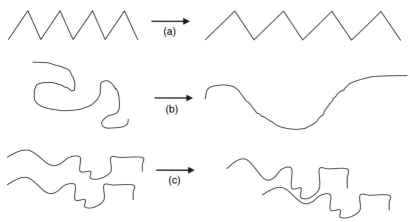

Figure 2.6 Deformation modes; a) bond bending; b) uncoiling; c) flow

and recovery will not occur. This type of deformation is generally called *viscous flow*. The chemistry of the polymer is a major contributor to the manner in which the chains interact. The physical properties that are measured in most tests result from the manner in which the polymer chains are aligned and move when at rest or in motion, either hot or cold.

Mechanical testing of plastic materials requires application of forces or energy to a material and observing the material's responses. These observed responses may be instantaneous, they may be time dependent, or they may result in a change of state of the material. The science that deals with the way materials deform when forces are applied is called **rheology**. There are two aspects of rheology:

1. The relationship of a material's response to force and deformation.

2. The relationship of how behavior is influenced by a material's structure and composition, that is, molecular structure, morphology, attributes, and so forth.

Before proceeding, some key terms need to be defined and discussed. These are:

Stress – The ratio of the quantity of force applied to a material to the cross sectional area of the material upon which the force is applied.

$$\text{Stress} = \frac{\text{Force}}{\text{Area}} \tag{2.3}$$

Greek letters σ (sigma), and τ (tau) are customarily used to denote stress.

Strain – This term is slightly more complex to perceive. It denotes the change of a quantifiable dimension of a material relative to the material's original dimension. That is, if a material that has a length, L_0, is stretched to a new length, L_{new}, the strain is the ratio of the new length to the original length.

$$\text{Strain} = \frac{L_{new} - L_0}{L_0} \qquad (2.4)$$

Greek letters γ (gamma) and ε (epsilon) are commonly used to denote strain.

Modulus – Modulus is a constant that expresses the measure of a specified material property such as elasticity or strength. Normally, modulus is an indication of a material's strength or stiffness. It is quantified by dividing stress by the strain when the strain is limited to elastic deformation only.

$$\text{Modulus} = \frac{\text{stress}}{\text{strain}} \qquad (2.5)$$

The capital letters E (Young's Modulus) and G (Shear Modulus) are normally used.

Viscosity – Viscosity is a time based quantity of a material's flow or resistance to flow. It is calculated by dividing the stress applied to a material by the time rate of change of strain.

$$\text{Viscosity} = \frac{\text{stress}}{\dfrac{\Delta \text{strain}}{\Delta \text{time}}} \qquad (2.6)$$

The Greek term η (eta) is the common notation for viscosity. The Greek symbol Δ (delta) means change. A simple means of understanding this concept is to consider speed or velocity as indicated by the speedometer on an automobile. The speedometer reads in miles per hour. More specifically it means the rate of change in distance per unit time or Δ miles / Δ time.

Plastics exhibit both solid (elastic) and liquid (viscous) properties simultaneously. If the material is in the solid state, the elastic component dominates. In the liquid state the viscous property dominates. In addition, depending on temperature, use conditions, time, and polymer attributes, either viscous or elastic property may be dominant over the other. The term for this behavior is **viscoelasticity**. Viscoelasticity is important in that it governs how a material will behave when processed. It also determines how and to what extent and under what conditions a material can be used. Finally, the mechanical and physical data generated when a plastic is tested result from its viscoelastic properties.

Equations 2.5 and 2.6 are called equations of state or constitutive equations. Explanations of their origin or derivation are beyond the scope of this primer. Viscosity and modulus are material constants. Basically, they define the make up of a material. From these equations or forms thereof one can begin to explain a lot about testing and consequently the behavior of material. Physical testing in the solid state primarily incorporates an analysis of the strength or durability of a material. Flow testing obviously involves measurements of viscosity in some form. These are two important points to remember as we will embark on an explanation of plastic testing.

In performing a test, there are four primary variables to consider. They are force, deformation, temperature, and time. At any given point, one of these variables can be arbitrary, that is, controlled by the tester. The others are observed or measured. The information generated from the test will reveal insight about the following:

• The strength of a material

• Longevity

• Shape retention

• Processing conditions

• Applicability

• Design criteria

• Setting specifications

• Classification

• Quality Control

In the following chapters attempts will be made to establish a sound foundation for understanding mechanical or solid state and flow property testing of materials using rheological concepts.

3   Mechanical Properties

3.1   Mechanical Testing in the Solid State

Mechanical properties are among the most important properties for material selection and end–use applications. Virtually all applications involve some type of material loading and responses. Consequently, such properties as modulus, tensile strength, and impact are essential for product design, material selection, and specifications. Mechanical behavior in general terms is concerned with the deformation that occurs under loading. The deformation will depend on the configuration of the specimen and the way in which the load is applied. Such considerations are left to the engineer and the testing scientist, who is concerned with predicting or evaluating the performance of a polymer. For this work a concern with a generalized constitutive relationship that relates stress to strain was used.

One of the simplest stress-strain relationships is termed *Hookes Law*, which relates stress from Eq. 2.1 in Chapter 2

$$\sigma = \frac{F}{A} \tag{3.1}$$

to strain in Eq. 2.2.

$$\gamma = \frac{\Delta L}{L} \tag{3.2}$$

This relationship yields the material quantity, modulus (Eq. 2.3). Therefore, Hooke's Law is

$$\sigma = E \cdot \gamma$$
$$\text{(i.e., Stress} = \text{Modulus} \times \text{Strain)} \tag{3.3}$$

Where E = Young's Modulus, σ = stress and γ = strain. This implies that stress is directly proportional to strain. This is indeed true for a perfect elastic solid. However, as mentioned earlier, plastic materials are viscoelastic and can become inelastic as a result

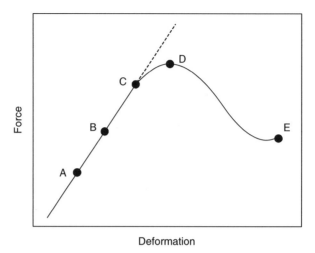

Figure 3.1 Typical stress – strain curve for polymers

of large deformations. Consequently, we must introduce some additional terminology to explain and quantify polymer behavior.

Figure 3.1 is an illustration of the stress-strain relationship of a typical polymer. First, at low strains, polymeric materials will indeed behave as an elastic solid and thus obey Hooke's Law. This is represented by the section of the curve labeled A – C. Equation 3.1 is an equation of a straight line. σ is the dependent variable along the y-axis and ε is the independent variable along the x-axis. Young's Modulus is the slope of the line. Hence, Hooke's Law is a linear relationship. Young's Modulus can be calculated at any point along the straight line portion of the curve.

It is important to note that modulus is a material parameter. That is, a specific material will have a modulus that is unique to it. It is related to the polymer attributes discussed in the preceding chapters. Consequently, any test that satisfies the conditions of linearity can successfully, accurately, and definitively be used to characterize a material.

The curve in Fig. 3.1 begins to deviate from linearity at point C. This point is the *proportional limit,* a quantity sometimes used for material characterization.

At point D, there is continued increase in strain without a corresponding increase in stress. This is called the *yield point.* The quantity *yield strength* is calculated by dividing the cross-sectional area of the sample into the force at the yield point (see Eq. 3.5).

As the figure shows, the sample continues to stretch until it reaches a maximum length allowed by the instrument or until rupture. This extension is called *elongation,* and is expressed in percent.

3.2   The Tensile Test (ASTM D638, ISO 527)

Tensile tests are used for grading, selecting, and design of material. The data is reported as Young's modulus, tensile strength, yield strength, elongation at yield, ultimate tensile strength, and elongation at break. Each value has specific meanings to the user.

Young's modulus is a true material parameter. As mentioned earlier, it is a unique material property that is related to polymer attributes. Young's Modulus is taken from the initial slope of the stress strain curve, where the stress is directly proportional to the strain.

The remaining properties are not true material parameters but contain a significant amount of information about the strength, utility, and behavior of a material. For example, knowing that a material has high tensile strength and low elongation means that the material is strong but brittle. Composites and highly filled materials have this quality. Materials with low tensile strength and high elongation are weak and ductile. This is typical of thermoplastic elastomers.

Sample shape and dimensions are set by the standardized ASTM method. Specimens are usually 165.1 mm (6.5 in.) long and 19.05 mm (0.75 in.) wide. Thickness is 3.2 mm (0.125 in.). ISO 527-7 refers to a thickness of 4 mm (0.157 in.). Specimens are characterized by a center section 12.7 mm (0.50 in.) wide and $\geq$ 50 mm (2 in.) in length. This gauge length section gives the samples it characteristic dog-bone shape. The rationale for this shape is that it localizes sample displacement in the gauge region for rigid samples.

The sample is placed in the grips of the testing machine, which pulls the sample apart at a prescribed rate. The force required to pull the sample apart and the amount of sample stretch are measured. These values along with the sample cross-sectional area in the gauge region are used to calculate tensile properties listed below (see Fig. 3.2 for an illustration of the tensile test).

Young's Modulus – A material constant taken from the slope of the stress strain curve in the linear portion. Stress is the measured force of displacement divided by the cross sectional area of the sample in the dog-bone region. The strain is generally measured by incorporation of a strain gage or an extensiometer. The strain must be localized to get a true measurement. The expression for calculating Young's Modulus is:

$$E = \frac{\sigma}{\gamma} \tag{3.4}$$

Where σ = Force/cross sectional area and $\gamma = \dfrac{\Delta L}{L}$

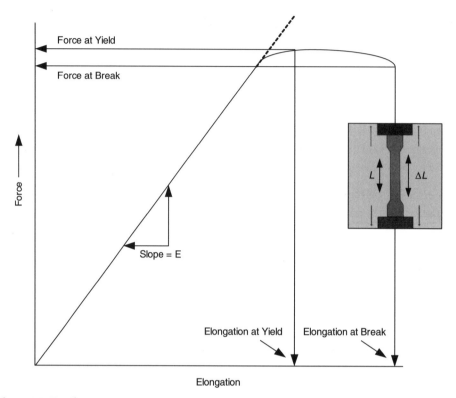

Figure 3.2 Tensile test curve, ASTM D698

Tensile strength – The point where elongation of the sample increases without a corresponding increase in force is considered the yield point. This force divided by the cross sectional area is by definition the tensile strength at yield.

$$TS = \frac{\text{Force at yield}}{\text{cross sectional area}} \tag{3.5}$$

Elongation at yield – The quantity of stretch at the yield point. It is expressed in percent.

$$EY = \frac{L - L_0}{L_0} \cdot 100 \tag{3.6}$$

Where L_0 = original sample length and L elongated sample length.

Ultimate Tensile Strength – The quantity of force applied when sample breaks divided by cross sectional area is the ultimate tensile strength of tensile strength at break.

$$UTS = \frac{\text{Force at break}}{\text{Cross sectional area}} \qquad (3.7)$$

Elongation at break – The quantity of stretch at the point of break expressed in percent.

3.3   Flexural Testing (ASTM D790, ISO 178)

Flexural testing is similar to tensile testing except that it is a three-point or a four-point bending test rather than a pull test. Flex testing is useful when stiffness is a concern. It is very useful in designing for structural applications. The values normally obtained from flexural tests are Young's modulus, flexural modulus, flexural strength, and secant. The value for modulus can be calculated using the linear portion of the stress strain curve from flexural tests (Tangent Modulus). This quantity is comparable to Young's Modulus from tensile tests. As the sample bends, the length of the sample increases, providing there is no slippage at the point at which the sample is supported. At small bending, slippage is nil. See Fig. 3.3 for an illustration of the flexural test.

Data normally reported from flexural testing are **Flexural strength, Flexural modulus, and Secant modulus.** The test is designed for the sample to fail before 5% strain is attained. If the maximum strain is reached without failure, the test is repeated with 10 times the strain rate.

Sample shape and dimensions are set by the standardized ASTM method. The molded specimen is usually 127 mm (5 in.) long 12.7 mm (0.5 in.) wide and 3.175 mm (0.125 in.) thick. The distance of the sample span between supports is 16 times greater than the sample thickness. Sheet samples can have similar dimensions if the thickness is ≤ 3.175 mm (0.125 in.). For thicker samples the width should not exceed one fourth of the support span. The sample is deflected at a rate of 0.254 mm (0.01 in.) per minute. The force required to deflect the sample to a strain of 5% is measured. These values along with the sample thickness are used to calculate tensile properties.

Young's Modulus is a material constant taken from the slope of the stress strain curve in the linear portion. Stress is the measured force of displacement divided by the cross sectional area of the sample. The strain is determined from the amount of deflection of a sample with a prescribed span.

$$E = \frac{\sigma}{\varepsilon} \qquad (3.8)$$

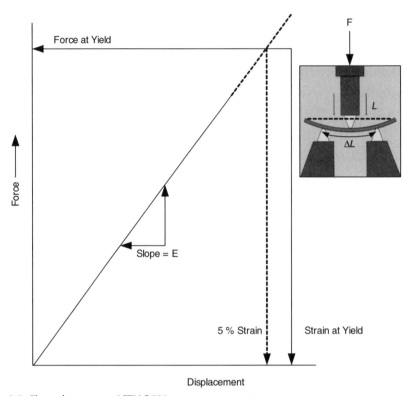

Figure 3.3 Flexural test curve, ASTM D790

Where σ = Force/cross sectional area and $\varepsilon = \dfrac{\Delta l}{l} = \dfrac{6\,D\,d}{L^2}$,

where D = deflections, L = span, d = depth (thickness).

Flexural Strength – Flexural strength may be calculated in one of three ways:
1. The maximum force the sample exerts on rupture is divided by the cross-sectional area;
2. Dividing the force at yield by the cross-sectional area;
3. If neither of the conditions in case 1 and 2 occurs, take the force at 5% strain to make the calculation.

Note: If condition 3 occurs, the method suggests increasing the test speed 10 fold or using a four-point bending test as described in ASTM D6272.

The general expression for calculating flexural strength is:

$$FS = \frac{\text{Force at rupture, yield, or 5\% strain}}{\text{Cross-sectional area}} \qquad (3.9)$$

Flexural Secant modulus – The force divided by the strain taken from any point on the stress-strain curve. The secant modulus ES is defined – using Hooke's Law – as the ratio of stress to strain, or the slope of a line drawn from the origin to a point on the stress-strain curve corresponding to the particular strain, typically between 0.2 and 7.0% for plastics.

3.4   Dynamic Mechanical Testing (ASTM D5279)

Dynamic mechanical tests (DMT) reveal the mechanical behavior of a plastic material over a range of temperatures and test rates. From this type of test one can get information about the effect of temperature on the modulus of a material. This test also provides information on properties such as softening point, glass transition temperature, window of use, and viscoelastic behavior.

The test is called dynamic because the sample is in constant motion during the entire test. The most common type of DMT is the torsion test or the twisting of a rectangular sample or rod. The sample is rigidly clamped at one end. The opposite end is twisted back and forth in an oscillatory motion. The magnitude of the twist is related to strain. The force required to twist the sample is related to the stress. Normally, there is a time lag between the signal to twist the sample and the actual response of the sample. The magnitude of the time lag or *phase shift* holds information related to viscoelasticity. There are two types of shear modulus obtained from this test. One is the storage or elastic modulus, G', the in-phase sample response. The other is the loss or viscous modulus, G'', the out-of-phase portion of the response. The ratio of G'' / G' is called tan δ (*tan delta*). When tan δ shows a peak, the sample is going through a thermal transition such as glass transition or melting.

Several new terms have been introduced in this section, which still need a definition. The variable G represents a shear modulus. From a physical perspective it is a pure elastic property like the Young's Modulus. Instead of an elongation deformation mode, here the mode is a shearing one. An example of a shearing mode can be illustrated by holding the palms of your hands together and sliding one hand forward while holding the other stationary. This sliding motion is shear. It is also illustrated in Fig. 2.2, where the polymer molecules are shown to slip relative each other. The shear modulus is a pure elastic material parameter and is defined by

$$G = \frac{\sigma}{\gamma} \tag{3.10}$$

Where σ is shear stress and γ is shear strain.

For dynamic testing we must further expand the definition of G to include the in-phase portion of the material's response, G', and the out-of-phase or time lag portion of the response, G''. The twisting motion eluded to earlier has two attributes:

First the magnitude of the twist is the angle of strain γ_0, for a rectangular beam

$$\gamma_0 = \left(\frac{t\,\theta}{2\,l}\right) \tag{3.11}$$

Where t is the sample thickness, l is the sample length, and θ = angle of twist.

The force required to achieve the strain is represented by the torque, M. By definition the stress is equivalent to the torque, M, over the cross-sectional area of the sample.

$$\sigma_0 = \frac{8\,M_0}{b\,t^2} \tag{3.12}$$

Where σ_0 is the stress amplitude and M_0 is the force applied to the sample in terms of torque, t = sample thickness, b = sample width.

Graphically the dynamic motion is best represented by two sine waves. One represents the incident signal to twist the sample and the second represents the sample response, which may be shifted over time (see Fig. 3.4).

Without offering complex mathematical derivations, the dynamic moduli are the following.

$$G' = \left(\frac{\sigma_0}{\gamma_0}\right)\cos\phi \tag{3.13}$$

and

$$G'' = \left(\frac{\sigma_0}{\gamma_0}\right)\sin\phi \tag{3.14}$$

Where G' is the elastic or storage modulus, G'' is the viscous or loss modulus, and ϕ is the phase shift (angle). When there is no phase shift ($\phi = 0$), $\sin\phi$ becomes zero, thus

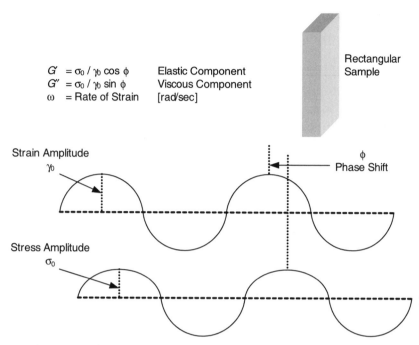

$G' = \sigma_0 / \gamma_0 \cos \phi$ Elastic Component

$G'' = \sigma_0 / \gamma_0 \sin \phi$ Viscous Component

ω = Rate of Strain   [rad/sec]

Rectangular Sample

Strain Amplitude
γ_0

Phase Shift ϕ

Stress Amplitude
σ_0

Figure 3.4 Dynamic mechanical test (solid state)

G'' becomes zero. At the same time, $\cos \phi = 1$. Therefore, when there is no phase shift the material does not have a viscous component. It is a pure elastic material. When the phase shift is completely out of phase, i.e., $\phi = 90^0$, the cosine becomes zero negating the elastic component. In this case the material is a pure viscous liquid.

As mentioned in Section 3.1, plastic materials have both a viscous and an elastic component. It is conceivable then that a dynamic test is an excellent method to evaluate the two components. Normally, when the elastic component is greater than the viscous component, the material will behave more as a solid. When the viscous component is greater, the material behaves more as a liquid. These relationships are both temperature- and test rate dependent as illustrated in Figs. 3.5 and 3.6.

Figure 3.6 is a solid state test of a rectangular sample at a constant rate of twist for polypropylene. The temperature is shown to vary between sub-ambient to the melt temperature of the material. G' and G'' are graphically illustrated. At low temperatures, G' is dominant and appears to be constant from very low temperatures until it reaches approx. 0 °C. G'' is orders of magnitude lower and also nearly constant. At zero degrees, there is a slight lowering of G' with a corresponding increase in G''. G'' peaks at about 5 °C. This peak represents the glass transition point for the material. The glass transition

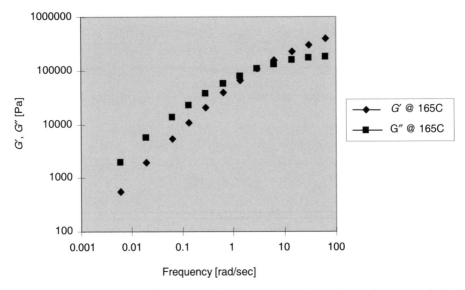

Figure 3.5 Dynamic data for polypropylene showing frequency dependence of viscous and elastic modulus

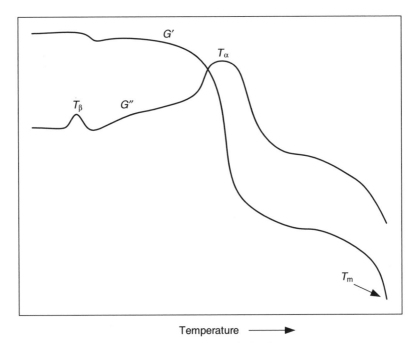

Figure 3.6 Molulus vs. temperature for an impact-modified polymer

for polymers is a point where sections of the molecule become free and more flexible. Below the glass transition point, the polymer is rigid and brittle. Above the glass transition point, the polymer is less rigid and ductile. A material has better impact properties above than below the transition point.

The next major transition is the melt point. The magnitude of this transition is significantly greater than any other transition one may observe. Discussion of the dynamic behavior of polymers in the melt state will be covered in the section on melt testing.

3.5   Impact Testing

Impact testing defines the ability of a material to absorb energy. Impact resistance is the ability of a material to resist breaking under a shock loading or the ability of a material to resist fracture under stress applied at high speed.

Most plastics break in a characteristic manner depending on the type of molecule, molecular components, method of fabrication, shape of the part, part complexity, orientation, and temperature. There are four types of impact failures:

*   **Yielding** – Evidence of permanent deformation without cracking

*   **Cracking** – Evidence of cracking or yielding without losing shape or integrity.

*   **Brittle** – Evidence of catastrophic failure without evidence of yielding. GP polystyrene is a brittle material.

*   **Ductile** – Evidence of definite yielding along with cracking. Polycarbonate is a ductile material.

3.5.1   Falling Dart Impact ASTM D5420 (No ISO Method)

The falling dart, drop impact, or Gardner impact test employs a free falling weight that strikes a plastic sample at rest (see Fig. 3.7). The energy required to rupture the sample is determined from the weight of the impacting device and the height from which it was dropped. The impact energy is calculated by multiplying the weight by the drop height. It is usually expressed in Joules (in.-lbf).

Gardner Impact   **ASTM D5420**

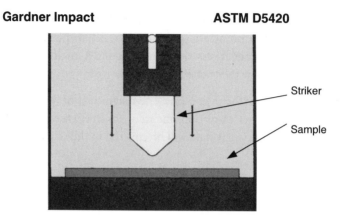

Striker

Sample

Figure 3.7 Falling dart (Gardner) impact test

GARDNER IMPACT TESTER

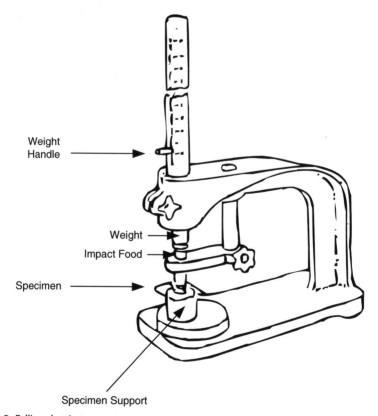

Weight
Handle

Weight

Impact Food

Specimen

Specimen Support

Figure 3.8 Falling dart instrument

The test is normally conducted by raising a weight to a desired height and allowing it to free-fall on a striker that is resting on a flat sample (see Fig. 3.8). The striker transfers the energy to the sample. The kinetic energy transferred by the falling weight at the instant of impact is equal to the energy used to raise the weight to the drop height. This is equivalent to the potential energy possessed by the weight as it is dropped. The mean failure height is determined from a statistically significant number of drops that will result in failing of 50% of the specimens. From this the mean failure height, the impact energy, is calculated (mass × gravity × height). This technique for determining the mean impact energy is called the **Bruceton Staircase Method** or Up-and-Down Method.

NOTE: Falling weight impact results are dependent on the geometry of the falling weight, striker, and support. In addition, it is dependent on the sample configuration and the preparation methods. As a result, impact testing should be used only to compare and rank materials. Impact values from this technique cannot be considered as absolute. The desired approach is to use equipment and test specimens that conform to the intended end use of the material. Data obtained from different geometries cannot be compared directly to each other.

3.5.2   Instrumented Impact Testing ASTM D3763 (ISO 6603.2)

Another method for determining impact rapidly gaining popularity is the instrumented impact test. In this method, sufficient energy is used to ensure sample failure with each strike, thus eliminating the need for multiple tests. A force-distance curve is generated and used to calculate impact energy by integrating the area under the force-distance curve. Instrumented impact tests have an advantage over falling-dart impact tests in that a quantitative value results from each test and it is possible to determine the mode of failure by examining the characteristics of the force-distance curve.

For the instrumented impact test a plunger is propelled at a speed of 200 m/min at a securely fastened sample. It should deliver sufficient energy to fracture the sample on impact. The available energy is set to achieve no more than a 20% slowdown in plunger speed after the plunger strikes the sample. The force required to pass through the sample is captured electronically and presented as a function of displacement (see Fig. 3.9).

Figure 3.9 illustrates the typical force displacement traces one should get for three types of plastic samples. Brittle failure is illustrated in Fig. 3.9a; it is characterized by a rapid increase in force followed by a sudden drop in energy within a short displacement. Ductile failure (Fig. 3.9b) is characterized by a less rapid increase in load with a clear indication of yielding before fracture. The third illustration, Figure 3.9c, appears to be a combination of the two. Ductile materials that are highly oriented are examples of materials exhibiting this type of behavior. Laminates and reinforced materials can also show similar curves when tested.

Energy curves can take many different forms and there are a number of characteristic curves instrumented impact testing will provide.

Impact properties of plastic materials are impact rate-dependent. In addition, sample geometry, preparation, thickness, and temperature can influence properties. Therefore, comparisons should be made using samples that are similar in geometry and prepared using similar techniques.

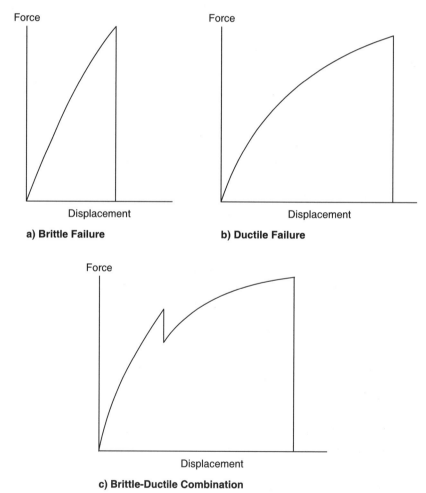

Figure 3.9 Typical force – distance curves generated by instrumented impact testing

3.5.3   Izod – Charpy Impact (ASTM D256, D4812, ISO 179)

This type of impact testing measures a sample's relative susceptibility to a pendulum-type of loading. The results are expressed in terms of the energy consumed by the pendulum to break the sample. Samples for this type of test are usually notched, because notching promotes brittle instead of ductile failure. The notch serves as a stress concentration area and therefore promotes crack propagation.

- **Izod Impact** – the test specimen is clamped vertically into position with the notch end facing the direction of the pendulum (see Fig. 3.10). Specimen dimensions as defined by ASTM are 50.8 mm (2 in.) long by 12.7 mm (0.5 in.) wide and 3.175 mm (0.125 in.) thick (sometimes 6.35 mm [0.25 in.] samples are used). The notch depth measures 2.54 mm (0.10 in.) and must be cut in a carefully prescribed manner. The pendulum is released, allowed to strike the sample and swing through. The sample must break completely in order for the data to be useful. An energy value is taken from the pendulum swing and divided by the sample thickness. Results are reported in Joules (in-lb/in. of notch).

- **Charpy Impact** – The only difference compared to the Izod test is that the sample is placed horizontally. The sample must break completely. Results are reported in Joules (ft-lbs).

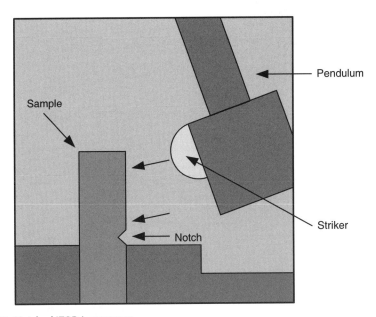

Figure 3.10 Notched IZOD impact test

3.6   Compression Tests – ASTM D695 (ISO 604)

Compression tests provide information about the mechanical properties of rigid plastics, including high modulus composites, when loaded in compression at low rates of straining and loading. Compressive properties are useful when a material is used under conditions similar to those in which the test is made. This type of test is useful for materials that fail by shattering under load. It is not particularly useful for ductile materials. However, compression testing is useful for determining permanent deformation (compression set) for thermoplastic elastomers.

This test is similar in concept to tensile and flexural tests except the deformation mode is compressive rather than stretching. Compressive properties include modulus of elasticity, yield stress, deformation beyond yield, and compressive strength. Modulus and yield values are determined in a similar fashion as presented in the sections on tensile and flexural tests. The most useful property obtained from this test is compressive strength. Compressive strength is determined by dividing the maximum load the sample carries by the original cross sectional area (Fig. 3.11).

Figure 3.11   Close-up of compression text fixture with prism sample

Test specimens are usually cylindrical or prism-shaped. Sample dimensions are preferably 12.7 mm (0.5 in.) diameter and 25.4 mm (1 in.) long. Other sample configurations are permissible; however, care should be taken with length to width or diameter ratios. If modulus of elasticity is desired, this ratio should range from 11 to 16 to one. The preferred rate of loading is 1.3 mm/min (0.05 in./min). For very ductile samples the speed may be increased to 5 to 6 mm/min (0.2 to 0.25 in./min) after yielding.

The following are the significant data obtained from compressive testing:

- **Compressive Strength** – Compressive strength is the maximum load the sample carries divided by the cross sectional area of the sample. It is expressed in megapascals or pounds-force per square inch.

- **Compressive Yield Strength** – Calculate the yield strength by dividing the load at yield by the cross sectional area of the sample. These values are reported in megapascals or pounds-force per square inch.

- **Modulus of Elasticity** – The elastic modulus is calculated by drawing a tangent line to the linear portion of load-deformation curve, selecting any point on the tangent line and dividing by the strain at that point. Modulus is reported in giga-pascals or pound-force per square inch.

Advantages of compressive testing:

- Useful for brittle materials experiencing load

- Flexibility in sample configuration

- Standardize test for high strength composites

Disadvantages of compressive testing:

- Results are dependent on test conditions and sample configuration

- Sample preparation can affect results

- Data will experience some scatter, therefore multiple test are required

- Not useful for ductile samples

3.7   Solid State Creep Test – ASTM D2990 (ISO 899)

A key issue with viscoelastic materials is that their properties are dependent on temperature, loading rate, and time. Therefore, short-duration tests cannot predict long-term behavior of a material. Tests must be conducted to measure the response of viscoelastic materials under stress and time. The solid state creep test was developed for this purpose. Data from creep tests are useful for determining the creep modulus and strength of materials and for the prediction of dimensional changes under long-term load. The information is useful for comparing materials, design of parts, and to characterize plastics for long-term use.

Creep testing can employ any of the test modes discussed in this section: tensile creep, flexural creep, and compressive creep. The test method consists of measuring the extension or compression as a function of time and time to rupture or failure of a sample subjected to a constant load under specified environmental conditions (Fig. 3.12).

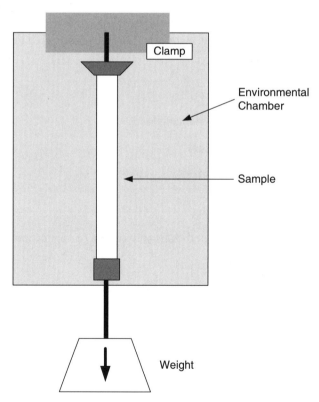

Figure 3.12 Typical creep test apparatus

When a load is applied, an instantaneous deformation will occur as a pure elastic response. The instantaneous deformation is followed by a rapidly decreasing deformation, called primary deformation. Primary deformation is followed by steady-state linear deformation, called secondary deformation. As the sample approaches failure, the deformation accelerates until fracture, which is called tertiary deformation. Figure 3.13 graphically illustrates the different stages of creep deformation.

Some advantages of creep testing are:

• An excellent indicator of long term properties

• Flexibility in testing modes

A key disadvantage of the creep test is that it is a long-term test. Some tests may take as much as 1000 hours or more to complete and gain meaningful data.

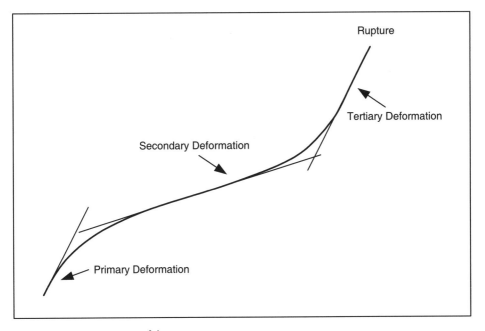

Figure 3.13 Various stages of the creep test

4   Thermal Testing

4.1   Introduction

Thermal analysis consists of techniques in which a sample's response to time and temperature are measured when exposed to a programmed temperature change. There are many tests that take on the form of thermal analysis. Issues regarding heating, cooling, melting crystallization, and degradation are the primary concerns. The key material properties are melting temperature (T_m), crystallization temperature (T_c), glass transition temperature (T_g), expansion, and contraction. These properties employ the following primary modes of tests:

- Differential Scanning Calorimetry (DSC)

- Thermogravimetric Analysis (TGA)

- Thermomechanical Analysis (TMA)

- Heat Distortion (HDT)

Coefficient of expansion, shrinkage, and thermal conductivity, or K Factor, are also key properties affecting processing. There are specialized tests for each one of these properties.

4.2   Heat Deflection Test (ASTM D648, ISO 75)

The heat deflection or distortion test (HDT), like the Flex test, is a 3-point bending test. This test measures the temperature at which a given sample will deform a specified amount under a prescribed load (Fig. 4.1). The sample dimensions are 50.8 mm × 12.7 mm × 6.35 mm (5 in. × 0.5 in. × 0.25 in.). The specified deflection is 0.25 mm (0.01 in.). Loads are defined as 0.455 MPa (66 psi) and 1.82 MPa (264 psi). HDT is normally used to define a material's temperature resistance and is therefore especially important to thermoformers as it serves as a guideline for setting mold temperatures.

The sample is placed in a three-point bending fixture with supports 101.6 mm (4 in.) apart. The force is applied to the thickness side of the sample rather than the width side, as with flexural tests. The assembly is placed in an oil bath. The sample is then preloaded with the prescribed load. The oil medium is heated at a rate of 2 °C/min. On heating the sample will soften and begin to deflect. When deflection reaches 0.25 mm (0.01 in.) the temperature is recorded and reported as the material's heat distortion temperature.

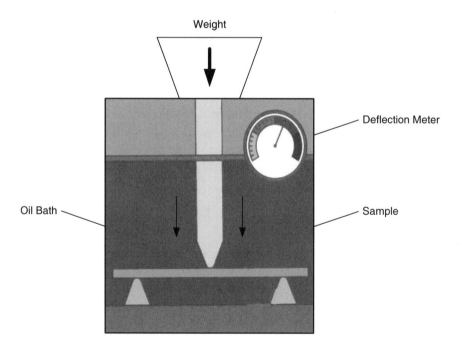

Figure 4.1 Heat deflection test

4.3   Vicat Softening (ASTM D1525, ISO 306)

The Vicat softening temperature is the temperature at which a flat needle will penetrate a sample a total of 1 mm under a given load and heating rate (50 or 120 °C per minute). This test is similar to HDT but its applicability is limited. It is primarily used for specific design or quality control purposes (see Fig. 4.2).

The sample is placed flat in an oil bath. The specimen has a minimum thickness and width of 12.7 mm (0.50 in.) and 3.05 mm (0.12 in.), respectively. Acceptable loads are 10 and 50 N, depending on sample type. Samples may be heated to 50 °C or 120 °C. The needle should have a blunt end with a surface area of 1 mm^2. The temperature at which the needle penetrates the sample 1 mm is recorded.

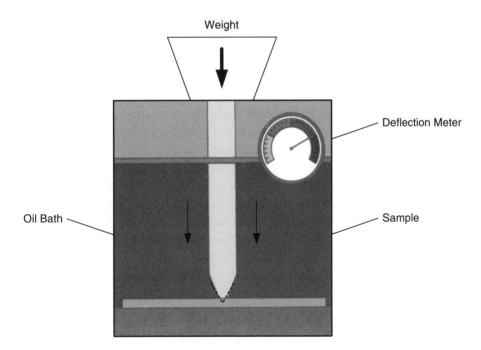

Figure 4.2 Vicat softening test

4.4   Differential Scanning Calorimetry, DSC (ASTM D3417, D3418)

In a differential scanning calorimetry test, the heat that flows into or out of a sample is measured while the temperature the sample is exposed to is programmed. The heat flow is a differential that is proportional to the temperature difference between the sample and a reference. When a sample goes through a transition such as from a solid to a melt, or vice versa, heat is either absorbed or emitted without a corresponding change in the temperature of the sample. The measured heat flow will show a peak in the curve.

A small amount of sample, usually 5 to 10 milligrams is sealed in a conductive pan. The pan, along with a reference pan, is placed inside a well-insulated oven (Fig. 4.3). The temperature in the oven is programmed to heat or cool in a prescribed manner And the energy required to heat or cool the sample is measured. At transition points, more or less energy is required, depending on whether the transition is endothermic (heat absorbing) or exothermic (heat emitting).

Figure 4.4 illustrates the typical data from DSC tests. As can be seen, there is substantial information to be gained from DSC tests about the thermal behavior of a material. Figure 4.4a is an illustration of an endothermic reaction such as a glass transition, showing the characteristics of the glass transition state. The peak in Fig. 4.4b illustrates the exothermic reaction, as the material transitions from an amorphous to a crystalline state as seen with polyethylene terephthalate. The crystalline melting point (Fig. 4.4c) is endothermic, that is, energy is absorbed. When the material degrades, the transition is endothermic, see Fig. 4.4d. Crystallization from the melt state is exothermic (Fig. 4.4e); therefore, energy is given off. Most chemical reactions are exothermic.

4.5   Thermogravimetric Analysis (TGA)

In this procedure, the weight of a sample is continuously monitored as the sample temperature increases. When a sample decomposes or degrades, normally the decomposition components volatilize off, resulting in a change in weight. A typical apparatus consists of a sample crucible, supported by an analytical balance inside a heating chamber. TGA is an effective way to measure the inert filler content of a sample.

Typical DSC Apparatus

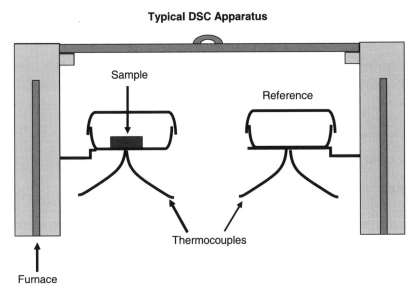

Figure 4.3 Differential scanning calorimeter apparatus

Typical polymer DSC thermograms

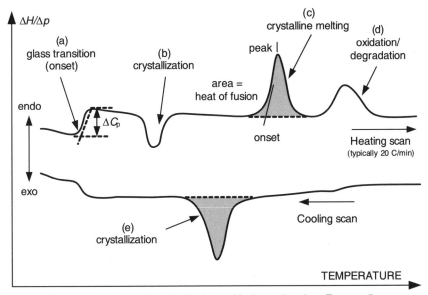

Figure 4.4 Representative endotherms and exotherms from DSC testing

4.6   Thermomechanical Analysis (TMA)

When a sample is heated, its dimensions change because of thermal expansion. TMA measures these changes using constant force on the sample. The TMA apparatus consist of

- A force transducer to control the force applied to the sample

- A position transducer to measure displacement

- A temperature controlled sample specimen

The sample geometry is defined and deformation occurs in a defined manner. Deformation modes include compression, tension, and three-point bending.

4.7   Thermal Conductivity, *K*-Factor (ASTM C177)

Thermal conductivity is defined as the rate at which heat is transferred by conduction through a unit cross sectional area of a material, when a temperature gradient exists across the material. The coefficient of conductivity, sometimes called the *K*-Factor, is expressed as the quantity of heat that passes through a unit cube of a substance in a given unit of time when the difference in temperature between the two faces is 1 °C.

$$K = \frac{Q\,t}{A\,(T_1 - T_2)} \tag{4.1}$$

Where K is the thermal conductivity; Q is the heat flow (BTU/hr); t is the thickness of specimen (in.); A is the area under test (in.^{2}); T_1 is the temperature of the hot side (°F); T_2 is the temperature of the cold side.

The procedure for determining the *K*-Factor is complex and will not be discussed here. The apparatus is referred to as guarded hot plate. Fortunately, *K*-Factors for most commercial grade materials have been determined and are readily available in handbooks.

4.8   Thermal Expansion (ASTM D696, ISO 3167)

The coefficient of thermal expansion is defined as the fractional change in length or volume of a material for a unit change of temperature (Fig. 4.5).

Expansion and contraction develop internal stresses and stress concentrations in materials that can lead to premature failure. Also, mold design and trimming stages must take into consideration a material's expansion-contraction characteristics. The thermal expansion coefficient is calculated as follows:

$$X = \frac{\Delta L}{L_0 \, \Delta T} \tag{4.2}$$

Where X is the thermal expansion coefficient (in./in./F); ΔL is the change in length of sample on heating or cooling (in.); L_0 is the original sample length at reference temperature; ΔT is the temperature between reference temperature and the temperature at which the measurement is made.

The test uses a quartz-tube dilatometer, a device for measuring changes in length, and a liquid bath to control the temperature. The test is started by mounting a specimen 50 mm (2 in.) to 127 mm (5 in.) long into the dilatometer. The dilatometer along with the measuring device is placed in the liquid bath. The temperature of the bath is varied as specified. The changes in length are measured.

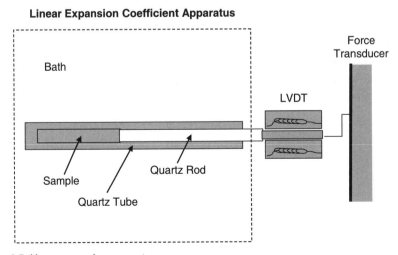

Linear Expansion Coefficient Apparatus

Figure 4.5 Linear expansion apparatus

4.9   Orientation, Shrinkage (ASTM D2732, ISO 11501, D2838)

ASTM D2732 determines the degree of unrestrained linear shrinkage at a given temperature of a specimen at least 0.763 mm (0.030 in.) thick. A sample of known dimensions is in a free shrink sample holder that should prevent the sample from floating and allow for the oil to contact all sides of the sample. The sample is immersed in an oil bath at a defined temperature for 10 seconds. The sample is then quickly immersed in a bath at room temperature for 5 seconds. Sample dimensions are measured and compared to dimensions prior to immersing into the hot bath. Data is reported in percent.

ASTM D2838 determines shrink tension of a restrained specimen of at least 0.763 mm (0.030 in.) thickness. The basic procedure is similar to ASTM D2732, except that the sample is restrained. A transducer is incorporated to measure the forces generated while shrinkage occurs. Results are reported in terms of force per unit of cross-sectional area, Pascals or psi.

4.10   Free Standing Orientation Test (ASTM D1204)

This test is widely used in thermoforming for sheet quality. A sample usually 254 mm (10 in.) by 254 mm (10 in.) is placed in a preheated oven heated to a desired temperature. The sample is free standing and coated with talc or a similar material to reduce friction. After allowing sufficient time for the sample to relax, it is removed and measured. The new dimensions are compared to the original dimensions. The data are reported in percent as well as any visual observations necessary.

5   Viscous Flow Properties

5.1   Introduction

Flow behavior of polymers is extremely important for the processor, material supplier, the end user, and the polymer scientists. This property determines whether a material can be used in a given manufacturing process; it establishes processing conditions; it determines if there are undesirable residual stresses; it classifies materials and aids in material development. As mentioned earlier, plastics are viscoelastic materials. That is, they have both a viscous (flow) property and an elastic (solid) property simultaneously. In the liquid state the viscous component is dominant.

Viscosity is one of the most easily measured and frequently used properties to characterize thermoplastic materials. Viscosity depends on the following factors:

- Flow Conditions
  Shear rate
  Temperature
  Pressure

- Material Property
  Chemical structure of the polymer
  Molecular weight and molecular weight distribution
  Long chain branching
  Nature and concentration of fillers, additives, and others

We will attempt to present a brief explanation of the effect of each of the above on viscosity. First, a short discourse on viscosity:

Viscosity is a property of a material that involves resistance to continuous deformation. Unlike elasticity, where the stress is proportional to the deformation, for viscosity the stress is related to the *rate* of deformation or strain. Thus, it is related to liquid flow rather than to solids. The simplest type of flow is represented by a linear relationship between deformation and stress. This is best represented by the proportionality expression

$$\sigma = \eta \, \dot{\gamma} \qquad\qquad (5.1)$$

Where η is the viscosity, σ is the stress, and $\dot{\gamma}$ is the strain rate. A material that behaves in this manner is called a *Newtonian* fluid. This relationship holds true for low molecular polymers or, at very low strain rates, for high molecular weight polymers. Newtonian viscosity is a material constant in that it does not depend on the rate or amount of strain but on the type of material.

For high molecular weight polymeric liquids, the flow behavior can become *non-Newtonian* at relatively low shear. In most cases, shear rates that are practical for study are in the region where non-Newtonian behavior is observed. For this case Eq. 5.1 must be modified, because in this case viscosity cannot be described by a single constant but rather by a strain rate dependent quantity. It is still convenient to use viscosity when characterizing materials. A typical viscosity – strain rate curve is illustrated in Fig. 5.1. There are two important aspects of the figure to consider.

1. At sufficiently low shear rates the viscosity approaches a limiting constant value called the zero shear viscosity. This value can be considered in much the same manner as the Newtonian viscosity. It is strongly dependent on molecular weight.

2. At higher shear rates the viscosity decreases. This type of behavior is called shear thinning. Molecular weight distribution influences this behavior. In addition, it best represents processing conditions.

It is obvious that a great deal of information can be obtained by measuring viscosity. These measurements have taken many different forms and used different methodologies. Those of significance in practical applications are the following:

Viscosity vs Shear Rate

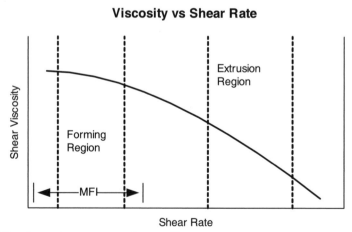

Shear Rate

Figure 5.1 Viscosity vs. shear rate curve from capillary test

- **Tube Flow (Fig. 5.2a)** – This flow is the one most often used for viscosity measurements. It occurs when the liquid is transported through a tube, pipe, or circular channel that is smaller in diameter than the reservoir containing the liquid. The small channel is referred to as a capillary.

- **Parallel Disk Flow (Fig. 5.2b)** – This is a torsional flow generated when the liquid is contained between two parallel plates and is sheared by the movement of one of the plates relative to the other.

- **Cone and Plate Flow (Fig. 5.2c)** – In this case, one plate is flat while the other has a cone shape. Similar to the parallel plates, this flow is torsional. However, in this case the flow field and stress fields are uniform. It represents a more precise measurement that is favored by rheologists

- **Concentric Cylinder (Couette) Flow (Fig. 5.2d)** – This is a drag flow generated by the rotation of either the inner or the outer cylinder. Couette flow generates higher torque compared to plates. Therefore, it is widely used for Newtonian and low-viscosity fluids.

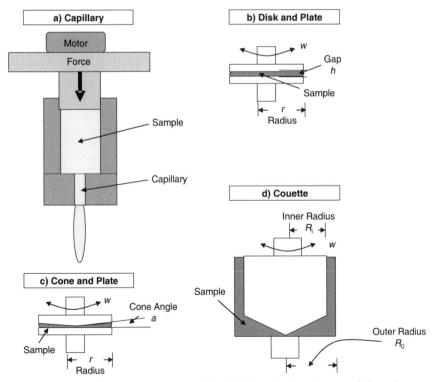

Figure 5.2 Rheological test geometries: a) capillary, b) disk and plate, c) cone and plate, d) couette

There are several other types of flow such as slit flow, annular flow, helical flow, etc., that are not widely practiced and will not be discussed in this work.

5.2   Melt Index Test (ASTM D1238, ISO 1133)

A familiar form of tube flow testing is the melt index test (MI, MFI, and MFR), which measures the rate of extrusion of a molten material through an orifice of specified length and diameter under a prescribed temperature and load (Fig. 5.3).

The melt index generally correlates inversely to the molecular weight of a material. High molecular weight materials have a low MI and low molecular weight materials have a high MI. Extrusion, thermoforming, and blow molding grades normally have low MI's. Injection molding and fiber grades have high MI's.

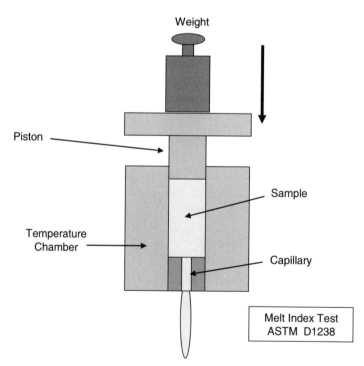

Figure 5.3 Typical melt indexer

$$MI = K \, \pi \, D_c^3 \, \frac{1}{\rho} \frac{dg}{dt} \tag{5.3a}$$

Where $K = 75 \, / \, 4$; $D_c =$ diameter of capillary; $\rho =$ melt density; $dg \, / \, dt =$ mass flow rate.

Or

$$MI = (426 \cdot L \cdot \rho) \cdot t \tag{5.3b}$$

Where $L =$ distance of piston travel, $\rho =$ density of melt, and $t =$ time. Or

$$10 \cdot \text{weight of sample after 1 minute extrusion} \tag{5.3c}$$

Melt flow quantities are used to designate grades of materials. Data are reported in grams of extrudate per 10 minutes or in decagrams per minute.

ASTM D1238 lists specific loads and temperatures for a given material. For example, polypropylene requires 230 °C and 2.16 kg load, while polycarbonate requires 300 °C and 1.2 kg load. The conditions are chosen to generate a relatively low shear rate. The shear rate should be reasonably low enough to be strongly influenced by molecular weight and less influenced by molecular weight distribution. Consequently, the melt flow index is routinely used to classify commercial materials according to their molecular weight.

5.3   Capillary Rheometry (ASTM D3595)

This technique is similar to the melt index test, except that the piston is motor-driven. Shear rates can be independently selected to be consistent with those experienced during processing. As a result, capillary data are used extensively for the design of polymer processing systems. Screw design companies and computerized process simulators rely heavily on capillary data.

In capillary testing a molten material is placed in a heated chamber. A motor-driven piston forces the material out of the chamber through a circular tube that is significantly smaller in diameter than the chamber. This forcing of the melt through a smaller orifice creates a shearing deformation of the material. The speed by which the melt is forced through the orifice is related to an apparent shear rate, γ_a. This leads to the following expression for shear rate:

$$\gamma_a = \frac{4\,Q}{\pi\,R^3} \tag{5.4}$$

Where Q is the volumetric flow rate and R is the radius of the capillary.

The stress is dependent on the force of the piston acting on the melt in the chamber. It is normally measured in terms of pressure. An appropriate term for the stress (σ) is

$$\sigma = \frac{R\,P}{2\,L} \tag{5.5}$$

Where P is the pressure, R is the radius of the capillary, and L is the capillary length. Note that the analysis for Eqs. 5.4 and 5.5 are independent of the rheological properties of the fluid, therefore they will hold true for both Newtonian and non-Newtonian fluids. Consequently, we can safely allow the equation for viscosity (Eq. 5.2) to apply in this case.

$$\eta_a = \frac{\sigma}{\gamma_a} = \frac{\pi\,R^4\,P}{8\,Q\,L} \tag{5.6}$$

If a log-log plot of the stress versus the shear rate yields a straight line, the fluid is considered Newtonian. The slope of the line is the viscosity. If the stress-shear rate relationship is non-linear, as often encountered in capillary testing, the viscosity is called *apparent viscosity*.

Considering Fig. 5.1 again for the illustration of a typical viscosity – shear rate relationship. Note that at low shear rates the viscosity is relatively high and appears to be approaching a constant value. If the viscosity becomes constant at very low shear rates, it is known as the zero shear viscosity and is similar to the Newtonian viscosity for the material. As the shear rate increases, the viscosity decreases monotonically. This is the shear thinning effect referred to earlier in this Chapter.

As mentioned previously, Newtonian viscosity is a material constant and is strongly dependent on molecular weight. In Fig. 5.4 is a typical plot of zero shear viscosity verses molecular weight. Note that initially the slope of the curve is unity. At a critical molecular weight the slope increases to a value of 3.4, showing a much stronger relationship between zero shear viscosity and molecular weight.

Shear thinning is the result of polymer molecules becoming more aligned in the capillary and therefore exhibiting less resistance to flow. Low molecular weight molecules will

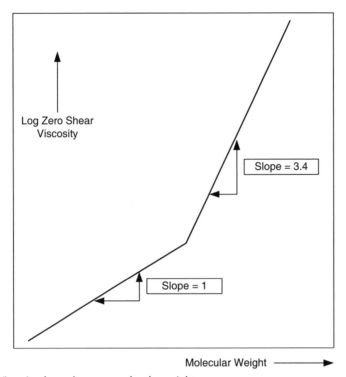

Figure 5.4 Viscosity dependence on molecular weight

align much easier and therefore will flow easier. If the melt has a significant amount of low molecular weight molecules in the distribution, shear thinning will be affected more than a system containing less low molecular weight molecules in the distribution. Therefore, it is safe to assume that shear thinning is dependent on molecular weight distribution.

Shear thinning can be quantified by calculating the slope of the shear stress versus shear rate curve in the high shear rate region. This slope is referred to as the *power law index*. The fluid is referred to as a *power law fluid*.

The shear rate in Eq. 5.6 is an apparent shear rate and is not the true shear rate of the system. To get the actual shear rate one must determine the slope over a range of shear rates and apply a correction factor. This correction factor is called the **Rabinowitch correction**. Specifically, the Rabinowitch correction allows for the true calculation of the shear rate at the wall. This is accomplished by conducting tests over a number of flow rates, tube lengths, and radii and plotting the log of the apparent shear rate (Eq. 5.4) versus the log of the shear stress (Eq. 5.4). The slope of the line is related to the shear rate at the wall. The expression for the Rabinowitch correction becomes

$$\gamma_w = \frac{3+b}{4}\gamma_a \tag{5.4a}$$

where

$$b = \frac{d\,(\log\gamma_a)}{d\,(\log\sigma_w)} \tag{5.4b}$$

Another correction that is meaningful for capillary testing is the Bagley correction. This correction accounts for entrance effects as the material enters the capillary. To obtain the Bagley correction one must measure flow properties at various capillary lengths. A plot is made of the pressure versus the length of the capillary to radius (L/R) ratio. Extrapolating L/R to zero gives the Bagley correction at the intercept (e) of the pressure axis. Thus, the true wall stress becomes

$$\sigma_w = \frac{P}{2}\frac{L}{R+e} \tag{5.4c}$$

Data without these corrections can be in error by as much as 20%. However, for comparison of similar batches, as in industrial quality control, it is often sufficient to use apparent non-corrected data.

5.4   Rotational Rheometry (ASTM D4440)

Rotational, or disk and plate rheology, testing has become more common and extremely important in recent years. Initially, the instruments designed for running these types of test were very expensive and difficult to operate. Data analysis and interpretation were equally complex and difficult. Recent developments in instrumentation and computer software have made rotational rheology testing routine and suitable for the quality control laboratory.

In disk and plate tests, one plate is held stationary while the second is set in motion. This motion imparts a shearing mode to the sample as illustrated in Fig. 5.5. There are two primary modes of motion one can apply:

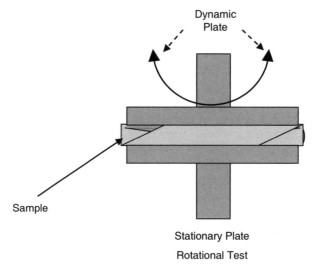

Figure 5.5 Rotational rheometry

Steady state test mode – a plate rotates steadily in a clockwise direction. Because of limitations in instrumentation this was the preferred test mode initially. Now it is primarily used for steady state creep studies. We will touch on this subject later in this chapter.

Oscillatory or dynamic test mode – this approach gained favor and became the most widely used technique because of its versatility and the amount of information one can get from a single test. This test yields viscous and elastic information simultaneously. It can be taken as a function of test rate (frequency), temperature, and time independently or together.

In Section 3.4 we developed the basic expressions for elastic and viscous modulus for a solid rectangular sample. Fortunately, the same basic equations apply here, except that the geometry of the sample is different. Here, the sample is a flat disk instead of a rectangle. However, as we are interested in fluids rather than solids we need to further refine the expression for modulus and introduce a **dynamic viscosity**.

Equations 3.13 and 3.14 for the elastic and viscous modulus are:

$$G' = \left(\frac{\sigma_0}{\gamma_0} \right) \cos \phi \qquad\qquad (5.7)$$

and

$$G'' = \left(\frac{\sigma_0}{\gamma_0}\right) \sin \phi \qquad (5.8)$$

We recall that viscosity is related to the rate of change of the material's shearing action. Consequently, if we introduce a velocity term into the equations, we can derive an acceptable expression for viscosity. Because we are oscillating the sample at a given frequency, f, we are exposing it to an angular velocity, ω. The angular velocity is

$$\omega = 2 \pi f \qquad (5.9)$$

By dividing Eqs. 5.7 and 5.8 by Eq. 5.9, we get two material functions $\eta'(\omega)$ (the dynamic viscosity) and $\eta''(\omega)$, both of which have the units of viscosity. Where

$$\eta'(\omega) = \frac{G''}{\omega} \qquad (5.10)$$

and

$$\eta''(\omega) = \frac{G'}{\omega} \qquad (5.11)$$

There is also a complex viscosity

$$\eta^* = \sqrt{(\eta'^2 + \eta''^2)} \qquad (5.12)$$

For fluids, η' is much greater than η'' and η^* is approximately equal to η'. Therefore, this author almost exclusively uses η' when discussing viscosity from dynamic tests. η' can be considered in much the same manner as apparent viscosity. In fact, the two are equal in magnitude when plotted on the same graph. Two scientists, W. P. Cox and E. H. Mertz, were the first to report that viscosities from capillary tests were equivalent. Therefore, this relationship has been called the **Cox-Mertz Rule**. It is conceivable that the same implications of viscosity on material behavior and composition apply to dynamic viscosity. The primary difference between the two viscosities is that dynamic viscosity is usually measured at lower rates than apparent viscosity.

There are two geometrical test modes for determining dynamic viscosity. One is parallel plate geometry; the second is cone and plate. Both will be discussed in the following sections.

5.4.1   Cone and Plate

In many instances, this is the ideal measuring system. It is very easy to clean, requires relatively small sample volumes, and with a little care it can be used on materials having a viscosity down to about ten times that of water (10 mPa-s) or even lower.

Cone and plate measuring systems are usually referred to by the diameter and the cone angle. Often cones are truncated. These types of cone are positioned such that the theoretical (missing) tip would touch the lower plate. By removing the tip of the cone, a more robust measuring system is produced (see Fig. 5.2c).

Because strain and shear rate are calculated using the angular displacement and the gap, it follows that the smaller the cone angle, the greater the error is likely to be in gap setting and hence in the results. By using a relatively large angle (4° or 5°), it becomes easier to get reproducibility of gap setting. The larger the cone angle, the more the shear rate across the gap starts to vary. Therefore, one should pay particularly close attention to cone angle when conducting cone and plate testing. For a 4° cone, the shear rate will vary by less than 0.5% across the gap, providing data with approx. 0.3% error. If a smaller cone angle is used, the operator-to-operator gap settings could easily introduce errors of over 5%, even by experienced operators, although the shear distribution error as such is small. Therefore, the larger angle gives a more acceptable error because it is a reproducible error. Fortunately, manufacturers of rotational rheometers have pre-selected the appropriate cone angle for users and recommend cone angle selection for the viscosity of the materials to be tested.

Because of the importance of correct positioning (often referred to as 'gap setting'), a cone and plate is not recommended when performing temperature sweeps, unless your rheometer is fitted with an automatic system for thermal expansion compensation.

You should also avoid using a cone if the sample you are testing contains particulate material. If the mean particle diameter is not some five to ten times smaller than the gap, the particles can 'jam' at the cone apex, resulting in 'noisy' data.

Materials with a high concentration of solids are also prone to being expelled from the gap under high shear rates, another reason to avoid the use of the cone.

Calculation of shear rate and shear stress from cone and plate tests are rather straight forward:

$$\text{Shear stress} = \frac{3}{2\,\pi\,r^3} \cdot M \tag{5.13}$$

$$\text{Shear rate} = \frac{1}{\theta} \cdot \omega \tag{5.14}$$

$$\text{Viscosity} = \frac{\text{Shear stress}}{\text{Shear rate}} \tag{5.15}$$

Where r is the radius of the cone, M is the torque, θ is the cone angle, and ω is the angular velocity.

5.4.2 Parallel Plates

The parallel plate (or plate-plate) system, like the cone and plate, is easy to clean and requires a small sample volume. It also has the advantage of being able to take pre-formed sample discs, which can be especially useful when working with polymers. It is not as sensitive to gap setting, because the separation between the plates (measured in mm) becomes part of the calculation (see Fig. 5.2b). Because gap setting can be arbitrary and controlled, parallel plate tests are ideally suited for testing samples using temperature gradients.

The main disadvantage of parallel plates lies in the fact that the shear rate produced varies across the sample. In this case, the average value for the shear rate is used.

Note also that the wider the gap, the more chance there is of forming a temperature gradient across the sample; therefore, it is important to surround the measuring system and sample with some form of thermal cover or oven.

Parallel plate systems are referred to by the diameter of their upper plate. The lower plate can be larger than or the same size as the upper plate.

When it is important to test samples at a known shear rate for critical comparison, the use of parallel plates is not recommended.

To make necessary calculations for plate and plate tests the following terms are useful:

$$\text{Shear stress} = \frac{3}{2 \pi r^3} \cdot M \tag{5.16}$$

$$\text{Shear rate} = \frac{3 r}{4} \cdot \frac{\omega}{h} \tag{5.17}$$

$$\text{Viscosity} = \frac{\text{Shear stress}}{\text{Shear rate}} \tag{5.18}$$

Where r is the radius of plate, M is the torque, ω is the angular velocity, and h is the spacing between plates (gap).

5.4.3 Concentric Cylinder – Couettes

Concentric Cylinder type measuring systems come in various forms such as coaxial cylinder, double gap, Mooney cell, etc., as shown in Fig. 5.2d. Standard coaxial cylinders are referred to by the diameter of the inner cup.

Couette measuring systems require relatively large sample volumes and are more difficult to clean. They usually have a large mass and large inertias and therefore can cause problems when performing high frequency measurements.

Their advantage comes from being able to work with low viscosity materials and mobile suspensions. Their large surface area gives them a greater sensitivity; therefore they will provide good data at low shear rates and viscosities.

The double gap measuring system has the largest surface area and is therefore ideal for low viscosity and low shear rate tests. It should be noted that the inertia of some double gap systems may severely limit the top working frequency in oscillatory testing. To calculate viscosity from coquette data use the following:

$$\text{Shear stress} = \frac{1}{2\, r_a^2\, H} \cdot M \tag{5.19}$$

$$\text{Shear rate} = \frac{2\, r_i^2\, r_o^2}{r_a^2\, (r_o^2 - r_i^2)} \cdot \omega \tag{5.20}$$

$$\text{Viscosity} = \frac{\text{Shear stress}}{\text{Shear rate}} \tag{5.21}$$

Where

$$r_a = \frac{r_i + r_o}{2}$$

r_i = inner radius
r_o = outer radius
H = height of cylinder

5.5   Solution Rheometry (ASTM 2857, ISO 1628)

Dilute solution testing was the initial method for measuring molecular characteristics for polymers. The technique is commonly used for approximating the average molecular weight of polymers in dilute solution. The measurement is usually made by using a glass viscometer, see Fig. 5.6. The viscosity of the fluid is determined by measuring the time in seconds for a given volume of fluid to flow a prescribed distance. The measurement is made using increasingly more dilute solutions. A viscosity is calculated for each concentration.

Several quantities are used in the determination. They are

- Relative viscosity $= \dfrac{\text{viscosity of solution}}{\text{viscosity of solvent}}$ (5.22)

- Specific viscosity $=$ relative viscosity $- 1$ (5.23)

- Reduced viscosity $= \dfrac{\text{specific viscosity}}{\text{concentration}}$ (5.24)

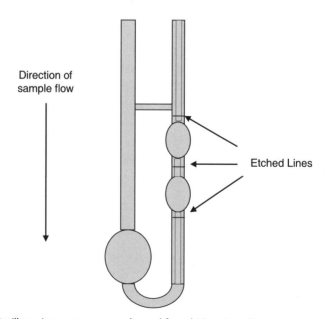

Direction of
sample flow

Etched Lines

Figure 5.6 Capillary viscometer commonly used for solution viscosity measurements

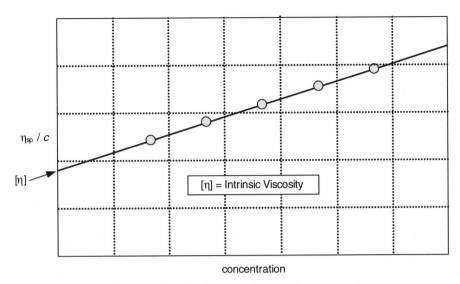

Figure 5.7 Illustration of solution viscosity data – specific viscosity vs. concentration

- Inherent viscosity = $\dfrac{\ln\,(\text{reduced viscosity})}{\text{concentration}}$ (5.25)

- Intrinsic viscosity = reduced viscosity at zero concentration
  = inherent viscosity at zero concentration   (5.26)

A plot is generated of the reduced viscosity and/or the inherent viscosity versus the concentration, Figure 5.7. The curve is extrapolated to zero concentration. The point at which the line intersects at zero concentration is called the **Intrinsic Viscosity** (IV). The intrinsic viscosity is directly related to weight average molecular weight. This quantity is used to classify materials for commercial applications.

5.6   Creep Test for Molten Polymers

A creep test consists of stressing a material and observing the change in deformation with time. From a creep test one can calculate the following:

$$\eta_0 = \frac{\tau}{\dfrac{d\gamma}{dt}}$$

$$J_r = \frac{\gamma_r}{\tau}$$

$$Je_0 = \frac{\gamma_{e0}}{\tau \lambda_0} = \eta_0 \, Je_0$$

Where η_0 = viscosity, J_r = recoverable compliance, Je_0 = equilibrium shear compliance, λ_0 = relaxation time, γ_r = recoverable strain, γ_{e0} = elastic deformation and τ = stress.

When the deformation rate is constant, steady state is reached. At this point, the time rate of change in strain, $d\gamma/dt$, is the strain rate. The ratio of strain rate to stress is viscosity. The elastic deformation can be obtained by extrapolating the steady state portion of the creep curve to zero time. By instantaneously relieving the stress, one can observe the amount of strain that is recovered and the amount that is permanent. The recovered strain is related to the elastic component. Figure 5.8 shows a graphical illustration of a typical creep curve showing the above relationships and variables.

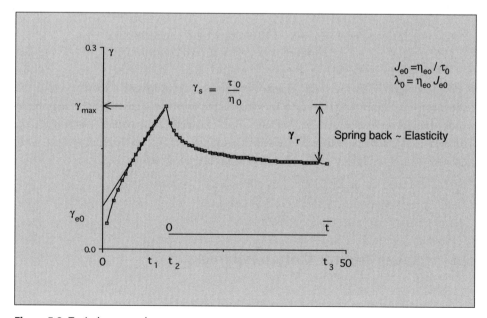

Figure 5.8 Typical creep and creep recovery curve

6 Quality in the Testing Laboratory

6.1 What is Quality?

In the late 1980s and early 1990s there was a major quality initiative within the American industry to educate and instill the concept of high quality and continuous improvement. This was triggered by the tremendous threat the industry in the United States was feeling from foreign entities. It led to such programs as the Malcolm Baldridge Award, ISO certification, and Six Sigma, to name a few. Many employees, including this author, were carted off to classrooms to learn and become indoctrinated to the new quality initiatives of their respective companies.

Over the years I have been fortunate enough to lecture on laboratory testing to individuals from all facets of the plastics industry. We end each session by having an open forum on quality, quality management, and quality assurance in testing and the testing laboratory. Our first task has always been to define the word *quality*. It has always amazed me how many different definitions we get, especially from individuals within the same company. It is also intriguing to hear the inputs from young bright freshman engineering students.

Before we can embark on the subject of creating quality in the laboratory, we must first complete a fundamental exercise. We must attempt to define what we mean by quality. Immediately, I am aware that there is perhaps no universal definition of quality in that it has so many meanings to so many different people. Therefore, we modify our thinking to define the attributes of quality instead, and how they should apply to a laboratory.

I would submit the following *attributes of quality* for the laboratory environment:

- A reasonable and acceptable **definition** of quality within the laboratory must be attained and applied. This concept should be mutually agreed upon and endorsed by all laboratory personnel and in particular by management. The emphasis is on mutual agreement by lab personnel and not a dictum from management. As a lab manager, I found this to be critical to the success of the lab.

- Attributes of quality should be **measurable**. Clearly establish means of evaluating and quantifying the process so that an accurate accounting can be maintained.

- It must be **controllable**. When things go wrong they can be fixed. More importantly, upsets can be prevented.

- All aspects should be **documented**. Access to history can prevent havoc in the future.

- **Continuous improvement** must be an essential part of the process. One should never be content with "this is the best we can do".

- It must be **universally accepted**. In this case, the universe is composed of those directly involved. By all means this includes the customer.

The above is by no means absolute, but I have found that commitment to this practice can create a quality environment that is palatable to all involved.

6.2   Quality Management

There are many types of laboratories where tests are conducted. There are research laboratories, development laboratories, service laboratories, and control laboratories. However, the ultimate laboratory is when the product, what ever it may be, is put in actual service. The challenge is to somehow predict, control, and understand the results in the ultimate laboratory in order to maintain an ongoing outlet for the product and the resulting profitable experience. It is evident that one of the most important aspects of maintaining this agreed level of quality is the place at which the data is generated, allowing the company to make a decision that the product is suitable for its customer. Ensuring that the information is accurate and precise is essential for success: this is where quality management comes in.

There are two key aspects of quality management: one is technical or systemic and the second is social or people. It has been relatively easy for organizations to implement, manage, and understand technical or systemic issues within an organization. These involve processes, systems, policies, and procedures. They represent the cornerstone of American industry. They are essential to the success and continuity of any organization.

Social issues are more difficult to understand and implement in a quality organization. These involve people, culture, and norms within and outside an organization. As the labor force has become more diverse, these issues have become more complex. Initially, all people involved were asked to fit in with the technical and systemic norms the organization has established. Those who cannot fit in must fit out. However, more companies are being asked or forced to look at their systems and modify them to suit the diverse issues within the organization. We will touch on some of these issues later in this section.

To achieve the objective of total quality management, we must fully understand our processes, our procedures, our people, and our customers' requirements. We most often characterize and evaluate our processes and products using measurements and test results made in the lab or on the floor. Therefore, we must fully understand each test procedure and how results are used and interpreted. Quality management in the lab requires that the lab understands its mission and be in control while providing results that are precise and accurate.

Below is a list of some caveats for total quality management in a laboratory:

- For each test performed, the lab must fully understand who the customer is and what their requirements for the test are.
- The lab is a customer of the supplier of the samples or materials to be tested. It is important for the supplier to understand the lab's requirements.
- Frequent communication between the lab and its suppliers and customers is essential.
- Laboratory personnel must be sufficiently trained.
- Complete understanding of each test is required.
- The lab must understand and implement statistics, precision, and accuracy.
- The lab must understand all of the quality issues, not just precision and accuracy.
- The lab must periodically conduct inter-laboratory comparisons within the lab and with suppliers and customers.
- The lab must understand that it is the cornerstone of total quality management within the organization.

6.3   Cultural Diversity and Quality

It is relatively easy to institute quality systems and policies. However, a difficult challenge to implementing a total quality program is the recruitment of personnel to "buy in" to the concepts of total quality and to commit to practicing them. Most laboratory personnel are well-trained and capable people that are committed to generating accurate and precise data. Most are eager to participate in any process designed to assure the reliability of the information they generate. Quality systems and practices should not be a top-down directive but an integral part of the culture of the laboratory. Perhaps a case history can best illustrate the value of buying in to a cultural change that resulted in a significant improvement in quality with a corresponding increase in productivity.

The laboratory was asked to provide physical property testing for a major plastics supplier. Many of the tests were similar to the tests covered in this work. The laboratory faced three major issues of concern. 1) There was a substantial backlog of outstanding work. 2) Quality was considered as not up to standards. 3) Morale was low. The laboratory was considered as a haven for malcontents and outcasts. A survey by external consultants revealed these issues.

The situation was further complicated by the fact that it was a culturally and gender diverse work force. The workers reflected the community from which the workforce was drawn.

A new manager was assigned to the laboratory with the charge to improve productivity and morale without negatively affecting quality. The new manager had been trained quality initiatives including personal and cultural diversity issues. After conducting individual interviews with each of the personnel, the new manager concluded the following:

- The personnel were well educated and very capable of high quality work.

- They felt that their abilities and contributions were not appreciated.

- They were committed to doing a good job and took pride in their work, but they felt that the company created obstacles that prevented them from doing their best.

- Their work styles were diverse and culturally biased. The biases were as follows:
  - Some viewed work as something very serious that should be orderly and well planned. They became upset with interruptions and changes in priorities and the lack of an overall plan.
  - Some thrived on uncertainty and constant change. This gave them an opportunity to provide a personal touch to their work and to improvise.
  - Others valued hard work and were very loyal to the cause as long as they were properly recognized.
  - Finally, there was a small group of perfectionists that would seek perfection at all cost.

However, there were two common traits that were consistent throughout the groups.

- They did not like the perception that they were characterized as a poor quality group. and
- They were very concerned and frustrated with the huge backlog of outstanding work.

The manager realized that the lab personnel themselves had defined the areas for "buy in". As a result it was an easy task to get them to agree on the internal mission of improving quality while increasing productivity.

The first major task was to develop a consensus definition of quality. They accepted definition of quality as "customer satisfaction". This was directly aimed at diffusing the poor quality perception of the laboratory. Each lab personnel was encouraged to initiate and maintain a dialog with their customers to determine their needs and what was considered as satisfactory results. The lab manager independently maintained contact with customers for the same purpose.

An extensive training program was initiated to train lab personnel to understand the tests for which they were responsible. Training covered theory, background, and practices for the various tests conducted in the laboratory. Professionals from the customer ranks were invited to present monthly seminars.

Internally, a statistical process control system was established. Customers were encouraged to review the data on a regular basis.

Personnel were given the option to choose the area within the laboratory they wished to work. Interestingly, they all opted to remain where they had been initially assigned, but with a new understanding and commitment that quality and productivity were their mission.

The issue of productivity was tackled by incorporating a work flow system to replace a priority system. The priority system artificially created a backlog of low-priority work that would never get addressed. The new system treated all work equally. The primary distinguishing factor between work requests was time. Emergencies or high-priority work had to meet specific criteria before work flow could be interrupted. A system for measuring and documenting work flow was included. The important factor in this change was that the personnel had input on the design and implementation of the new system.

The third major change was that the lab workers were empowered to develop their own methodology for implementation of the changes within their subgroups. This gave each the opportunity to satisfy their desire to comply with their personal and cultural demands while contributing to the primary objective.

The net result from the above effort was a laboratory that increased productivity by 200% while accomplishing a corresponding increase in quality. There were some skeptics that said this could not be done. But it was accomplished by including personnel as an integral part of the quality process.

The key ingredients for success were *inclusion* of personnel in developing the quality system and *empowering* them to implement the system while maintaining a non-intimidating *measuring* system.

6.4   Accuracy, Precision, and Bias

Shooting at a target, accuracy would typically be considered as hitting the bull's eye or at least come in very close proximity to it each time. Likewise, shooting at the target a number of times and hitting the bull's eye consistently, this would not only be accurate but precise as well (Fig. 6.1a). If the shots consistently miss the bull's eye, but each shot is in close proximity with one another, this would be considered inaccurate but precise or biased (Fig. 6.1b). If the bull's eye is missed and the shots are widely scattered, this would not only be inaccurate but imprecise as well (Fig. 6.1c). These are the three scenarios to consider when evaluating test data in laboratories.

Accuracy is sometimes arbitrary or we may not know exactly what the bull's eye is. Hence, ASTM advises us to consider accuracy as two components, precision and bias. Statistics will assist us in evaluating and understanding the precision and bias of our data.

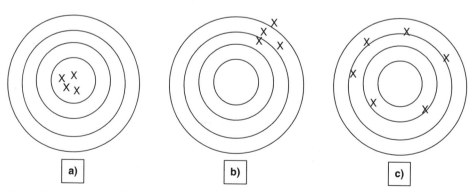

Figure 6.1 Accuracy and bias

6.5   Review of Basic Statistics

Control evaluation and improvement in a laboratory require a good understanding of statistics and its applications. Following are basic concepts of statistics that hopefully will be helpful. All experimental measures are variable, provided the experimental device is sensitive enough to detect variability. Unless measures show some variability, it is difficult to estimate the reliability of the measured difference. Therefore, it is desirable for repetitive measurements to vary but that the magnitude of variation to be small. An understanding of variability and how to deal with it is important. *Statistics* give us a tool to deal with variability.

Statistics is the study of methods of collecting, summarizing, presenting, analyzing, and interpreting numbers. Because it involves numbers, statistics can be applied anywhere where numbers are generated. It is clear that statistics contribute nothing to scientific accomplishment, test method, or product functionality. However, when the experimental measurements and observations have been recorded, statistical methods are used to reduce them to a simple form, which represents most accurately the phenomena studied.

Sampling for data collection is concerned with answering two questions:

How?
How many?

This means that we are interested in the *population*, that is, in all the numbers of interest and in the *sample*, that is, a representative of the population. A sample is selected in such a way that every member of the population has an equal chance of being selected.

The data collected from a number of samples must be analyzed to estimate the true values of the parameters of the population. Statistics is a tool with which this estimate can be made. First, it is recommended to present the data in a *histogram*. A histogram is a graphical method for illustrating the distribution of a set of data. It is a method to visually assess the precision of a data set.

The data will usually have a central tendency, that is, there is a typical number that indicates where the numbers tend to cluster. The common measures of central tendency are:

Mean *(average)* – The sum of a set of observations divided by the number of observations

$$\text{Average} = \frac{1 + 2 + 3 + 4 + 5}{5} = \frac{15}{5} = 3$$

Symbol – *X*bar or $\overline{X}$

Median – The middle value of a set of numbers arranged in ascending order. If the data set is even then the average of the two middle values is taken.

$$1, 2, \underline{3}, 4, 5 - \qquad \text{Median} = 3$$

$$1, 2, \underline{3, 4}, 5, 6 - \qquad \text{Median} = \frac{3 + 4}{2} = 3.5$$

Mode – The value of the observation that occurs most often.

$$1, 2, \underline{3}, 4, \underline{3}, 5, \underline{3} - \quad \text{Mode} = 3$$

Measure of the dispersion of the data requires determining the data range, standard deviation, variance, and coefficient of variance.

The **range** of the data is the difference between the largest and smallest value in a set.

$$1, 2, 3, 4, 5 - \qquad \text{Range} = 5 - 1 = 4$$

A measure of the variability of the data set is the **standard deviation**. Standard deviation is a statistic term used as a measure of the dispersion or variation in a distribution. It is a value that indicates how much the data varies above and below the average value of the set.

Standard deviation equals the square root of the sum of each of the values in the data set subtracted from the average of the set. This expression is squared and then divided by the number of data points less 1.

$$s = \sqrt{\frac{\left(\sum X - \dfrac{\sum X}{n} \right)^2}{n - 1}} \tag{6.1}$$

The square of the standard deviation is the *variance* of the set.

Another useful term is the **coefficient of variation** (*COV*). COV is simply the standard deviation divided by the average, usually expressed in percentage. It is useful for comparing data sets that have widely differing means. A COV greater than 30% indicates that there is a lot of scatter in the data and that the mean may be unreliable. A COV of 5% indicates that the data is very precise. An acceptable COV is approx. 15% or less.

6.6   Reasons for Data Variability

Listed below are some of the reasons data will vary:

*   Sample – Plastics are inherently inhomogeneous in structure. Various additives, fillers, and modifiers are added that contribute to more inhomogeniety. Processing plastics may result in degradation, orientation, and contamination. These variables are difficult to control; hence they are a common source of variability.

*   Sample preparation – Preparing plastic samples for testing requires strict adherence to prescribed methods.

    – When samples are molded, their properties may vary with mold design, molding conditions, and gating.

    – When the samples are cut, defects in the cutting tool, dull cutting edges, and rate of cutting can cause variation.

    – The age of the sample and the storage conditions of the sample can affect data.

*   Variations in sample dimensions will affect results, especially if the dimensions do not enter into the calculations.

*   Some additional sources of variation are:

    – Speed of testing
    – Testing temperature and humidity
    – Operator differences in technique, experience, and training
    – Instrumentation condition
    – Computational techniques

6.7   Statistical Process Control (SPC)

In order to maintain the desired level of accuracy, precision, and credibility, the lab must ensure that all tests are performed while the process is in control. A recommended technique for evaluating the performance of a laboratory is to develop and maintain process *control charts*. Practically anything in the laboratory that can be quantified can be charted. Control charting is a graphical technique, which plots results of samples and other data to monitor the stability of a process. It uses process capability rules to determine control limits.

Some benefits of control charts are:

- Control charts are simple and effective tools. They lend themselves to being maintained at the work station by the operator. They give the people closest to the operation reliable information on when action should be taken and when action should not be taken.

- Performance can be assessed at a glance, creating reliability between the customer and the data producer.

- When the process is in control it can produce further improvements to reduce variation.

- Subtle changes can be identified early and corrections can be applied.

- Control charting provides a common language for communication about the performance of a process.

The most common variables charted are the average value of a data set and the range between successive averages of data sets. These are called the Xbar and R charts. Xbar measures the central location of a subset of the sample population. The R chart measures the spread of the sample.

Xbar is plotted against a central point established for the data population. The expectation is that Xbar will fluctuate randomly about the central value. The random fluctuation should remain within a statistically determined upper and lower limit. These limits are determined from the standard deviation and confidence limits of the data. Any modern statistics text provides tables for determining upper and lower control limits.

The R chart, or moving range, is likewise plotted against a central range established for the data population. Upper and lower control limits are determined using the average range of the sample population and a constant taken from a statistics table.

The following is a case history illustrating the use and value of SPC charting. The test is a rotational rheometry test consisting of a frequency sweep using parallel plates at room temperature (see Section 3.4.2 for a description of parallel plate testing).

Table 6.1 is a tabulation of test data using a silicon standard. The variable is the frequency when $G' = G''$. The tests consist of a frequency sweep using parallel plates at room temperature. The control chart is shown in Fig. 6.2a.

Table 6.1 Sample Data for SPC Chart

Sample	Crossover frequency			Average
1	3.14	3.94	4.08	3.72
2	3.91	4.26	3.64	3.94
3	3.96	4.01	3.88	3.95
4	3.96	3.93	3.90	3.93
5	3.96	3.95	3.90	3.94
6	3.97	3.91	3.77	3.88
7	3.91	3.80	3.83	3.85
8	3.90	3.83	3.97	3.90
9	3.89	3.81	3.86	3.85
10	3.80	3.70	3.78	3.76

Ocillatory Test - Silicone Standard

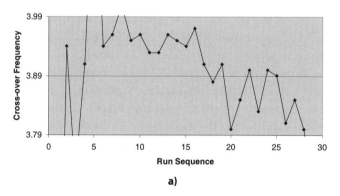

a)

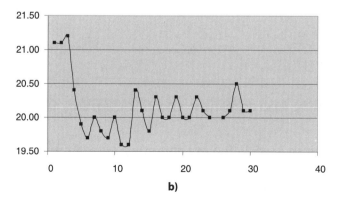

b)

Figure 6.2 a) SPC chart for silicone standard dynamic mechanical test
b) Chart showing room temperature at time of each test

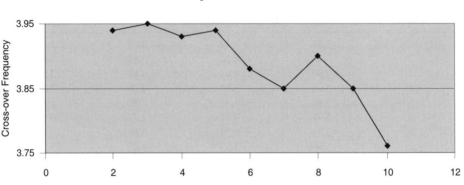

Figure 6.3 SPC chart for silicone standard with temperature control

The SPC chart shows that the laboratory is out of control as some of the data is completely out of range. In addition, there are eleven consecutive data points above the central point. This is followed by the majority of the remaining data points below the central point. Both of these conditions are also indicators of out-of-control conditions. After troubleshooting the process it was found that the temperature in the room was fluctuating. Figure 6.2b is a plot of room temperature taken at the time of each test. The data was inversely tracking room temperature. Because there was no effective way to control the room temperature, the temperature of the test was increased to 30 °C, The instrument is capable of maintaining control of this temperature internally and is independent of ambient conditions.

Figure 6.3 shows a new control chart with more precise temperature control. The process appears to be more in control.

6.8   Quality Accreditation and Sanctioning Organizations

6.8.1   American Association of Laboratory Accreditation (A2LA)

The American Association for Laboratory Accreditation (A2LA) is a non-profit, nongovernmental, public service, membership organization that operates laboratory accreditation systems. Accreditation is defined as recognition of the competency of a laboratory to perform specific tests or calibrations. Accreditation is available to any type of testing or calibration laboratory, private or government. A2LA was formed in 1978 to develop, manage, and recognize competent laboratories. Accreditation is available for virtually all types of tests, calibrations, measurements, and observations that are reproducible and properly documented.

Users of accredited laboratories are required to obtain the Scope(s) of Accreditation from any accredited laboratory or from A2LA. The Scope(s) of Accreditation identifies the specific tests, types of tests, or calibration capability for which the laboratory is accredited.

For tests, the scope of accreditation is normally identified in terms of standard test methods prepared by national, international, and professional standards writing bodies. Laboratories are expected to be competent in the use of the current version within one year of the date of publication of the standard test method.

For calibrations, the scope of accreditation is described typically in terms of the measurement parameter, range of measurement, and best attainable uncertainties.

The general requirements for A2LA accreditation are the international standards, ISO/IEC 17025-1999, *General Requirements for the Competence of Testing and Calibration Laboratories* (see Section 6.8.1.2). Additional program requirements for specific fields (e.g., calibration, environmental testing) or specific programs complement these general requirements. In effect, A2LA accreditation attests that a laboratory has demonstrated that:

- It is competent to perform specific tests, types of tests, calibrations, or types of calibrations listed on its Scope(s) of Accreditation.

- Its quality system addresses and conforms to all elements of ISO/IEC 17025-1999 (and, as a result ISO 9001-1994 or ISO 9002-1994), is documented per ISO/IEC 17025, and is fully operational.

- It is operating in accordance with its quality system.

- It conforms to any additional requirements of A2LA or specific fields or programs necessary to meet particular user needs.

6.8.1.1 Conditions for Accreditation

In order to attain and maintain accreditation, laboratories must comply with the Conditions for Accreditation published by A2LA. This document is available at the

A2LA website, www.A2LA.org, or from A2LA Headquarters. In order to apply, the applicant's laboratory's Authorized Representative, must agree to the conditions for accreditation and must attest that all statements made on the application are correct to the best of his/her knowledge and belief. An accredited laboratory's Authorized Representative is responsible for ensuring that all of the relevant conditions for accreditation are met. During the on-site assessment, the assessor will determine that the Authorized Representative and a deputy are knowledgeable about the accreditation requirements and they will require that the Authorized Representative and a deputy sign a statement that the Conditions for Accreditation will be upheld.

6.8.1.2 A2LA Accreditation Process

A laboratory applies for accreditation by obtaining the application package (available from A2LA headquarters or the A2LA website www.A2LA.org) and completing appropriate application sheets and relevant checklists. All applicants must agree to a set of conditions for accreditation and provide detailed supporting information, including:

- Proposed scope of testing or calibration, testing or calibration technologies, methods and relevant standards, and measurement uncertainty if applicable

- Quality manual

- Organizational structure

- Proficiency testing results

6.8.2   International Standards Organization Sanctioning (ISO/IEC 17025)

ISO/IEC 17025 – *General Requirements for the Competence of Calibration and Testing Laboratories* was published by ISO in December 1999. It is the result of a joint partnership between the International Organization for Standardization and the International Electrotechnical Commission.

ISO 17025 was developed specifically to give guidance to laboratories on both quality management and the technical requirements for proper operation. This standard can be considered the technical compliment to ISO 9000. Consequently, any organization that satisfies the requirements of ISO 17025 also meets ISO 9000 requirements.

While the ISO 9000 requirements are generic and are intended to be applicable to any type of organization, the ISO 17025 requirements are specific to testing and calibration laboratories. This standard addresses issues such as: the technical competence of personnel, ethical behavior of staff, use of well-defined test and calibration procedures, participation in proficiency testing, and contents of test reports and certificates.

Another reason for development of the standard was to harmonize laboratory accreditation and acceptance of test data worldwide. All participating countries will be required to accept the test results performed by accredited members of these other countries.

Reference Materials

1. Fred W. Billmeyer, *Polymer Science,* Interscience, NY, 1966

2. *American Heritage Dictionary of the English Language,* Houghton Mifflin CO., Boston, MA,1981

3. John D. Ferry, *Viscoelastic Properties of Plastics,* John Wiley and Sons, NY, 1980

4. I. M. Ward, *Mechanical Properties of Plastics,* 2nd edition, John Wiley and Sons, NY, 1985

5. John M. Dealy, Kurt F. Wissbrun, *Melt Rheology and its Role in Processing,* Chapman and Hall, NY, 1995

6. Vishu Shah, *Handbook of Plastic Testing Technology,* 2nd Edition, John Wiley and Sons, NY, 1998

7. Stephen Burke Driscoll, *The Basics of Testing Plastics,* ASTM Manual Series, 1998

8. Donald Hylton, *Material Testing and its Relationship to Thermoforming,* DCHylton Consulting, 2002

9. Tony Whelan, John Brydson, *The Kayeness Practical Rheology Handbook,* Kayness Inc., 1991

10. Gebhard Schramn, *A Practical Approach to Rheology and Rheometry,* Haake, 1994

12. *Quality Principles,* Exxon Chemical Company, 1989

13. Earl Burch, Frank Rudisill, *Quality Management in the Laboratory,* Professional Development, Clemson University, 1995

14. W. L. Gore, *Statistical Methods,* Interscience, NY, 1956

15. *Practical Rheology for the Engineer,* Course Handour, E. A. Collins, E. A. Collins Consulting

16. Elsie Y. Cross, *Managing Diversity: Valuing Differences for Organizational Effectiveness,* Elsie Y. Cross and Associates, 1988

17. *Annual Book of ASTM Standards,* Volume 8.01, 2002

18. *Annual Book of ASTM Standards,* Volume 8.02, 2002

19. *Annual Book of ASTM Standards,* Volume 8.03, 2002

20. *Annual Book of ASTM Standards,* Volume 8.04, 2002

21. American Association of Laboratory Accreditation, www.A2LA.org

22. International Standards Organization (ISO/IEC 17025)

Appendix 1

Partial List of Plastic Testing Instrumentation

Supplier	Web Site	Product, Service
Atlas Material Testing Technology, LLC	www.atlas-mts.com	Environmental, flammability, analytical
Bohlin Instruments, Inc.	www.bohlinusa.com	Rotational Rheometers, Viscometers
C. W. Brabender Instruments, Inc.	www.cwbrabender.com	Flow Testing,
Ceast, USA, Inc.	www.ceast.com	Izod, Instrumented Impact, MFI, Tensile, HDT Capillary
Clinton Instruments Company	www.clintoninstrument.com	Electrical
Custom Scientific	www.csi-instruments.com	MFR, Impact, Shrinkage, HDT
Dynisco, LLC	www.dynisco.com	MFR, Impact, HDT
Goettfert	www.goettfert.com	MFR, Capillary, Tensile
HunterLab	www.hunterlab.com	Color
Impact Analytical	www.impactanalytical.com	Analytical, Thermal Analysis
Instron Corporation	www.instron.com	Tensile, Impact, Capillary
Ircon Inc.	www.ircon.com	Temperature
Linear Laboratories	www.linearlabs.com	Temperature
Minolta Corporation	www.minolta.com	Color
MTS Systems Corporation	www.mts.com	Mechanical, Impact, Creep
Paul M. Gardner Company, Inc.	www.gardco.com	viscosity, Impact
Perkin-Elmer LLC	www.Perkinelmer.com	Thermal, Analytical
Q-Panel Lab Products	www.q-panel.com	Environmental
Raytek, Inc.	www.raytek.com	Temperature
Rheometric Scientific, Inc.	www.de.rheosci.com	Rotational, Impact

Supplier	Web Site	Product, Service
TA Instruments	www.tainst.com	Thermal Analysis, Rheometers
Thermo Haake	www.thermo.com	Rotational, Thermal
Tinius Olsen Testing Machines Company, Inc.	www.tiniusolsen.com	MFR, Capillary, Impact, Tensile
Zwick, USA	www.zwick.com	Tensile, Capillary, Impact

Appendix 2

Partial List of Testing Laboratories

OCM Test Laboratories	www.ocmtestlabs.com	Testing Services
Plastics Technology Laboratories, Inc.	www.ptli.com	Testing Service
PlastiScience, LLC	www.plastiscience.com	Testing Services
Plastics Technology Laboratories, Inc. /	www.ptli.com	Testing Service
L. J. Broutmann and Associates	www.broutman.com	Testing Services
Adtech Systems Research Inc.	www.erinet.com\adtech	Testing Services

Appendix 3

Nomenclature

A	Area
COV	Coefficient of variance
D	Diameter
E	Young's modulus
EY	Elongation at yield
FS	Flexural strength
G	Shear modulus
G'	Elastic or storage modulus
G''	Viscous or loss modulus
H	Height
J	Compliance
K	Thermal conductivity
L	Length
M	Torque
M_0	Tirque amplitude
MFI	Melt flow index
MFR	Melt flow rate
MI	Melt index
M_n	Number average molecular weight
M_w	Weight average molecular weight
P	Pressure
Q	Volumetric flow rate
T	Temperature
TS	Tensile strength
UTS	Ultimate tensile strength
X	Thermal expansion coefficient
X	Average
$\overline{X}$ (Xbar)	Average
d	Depth
e	Error correction
f	Frequency

g	Acceleration of gravity
h	Gap displacement
r	Radius
s	Standard deviation
t	Time

Greek Letters

δ	delta	Mechanical loss angle
Δ	delta	Change in
ε	epsilon	Strain
ε^*	epsilon	Complex viscosity
η	eta	Viscosity
η''	eta double prime	imaginary or elastic viscosity
η'	eta prime	Real viscosity
γ	gamma	Strain
γ_0	gamma	Strain amplitude
γ_a	gamma	Apparent shear rate
γ_w	gamma	Shear rate at wall
$\dot{\gamma}$	gamma dot	Strain rate
λ	lambda	Relaxation time
ω	omega	Angular velocity
φ, ϕ	phi	Phase shift
π	pi	3.14
ρ	rho	Density
σ	sigma	Stress
σ_0	sigma	Stress amplitude
Σ	sigma	Summation
τ	tau	Stress
θ	theta	Strain angle

Appendix 4

Referenced ASTM Methods (in order of appearance in text)

ASTM	Title	ISO Equivalent
D638	Test method for tensile properties of plastics	527
D790	Test method for flexural properties of unreinforced and reinforced plastics and electrical insulating materials	178
D5279	Test method for plastics: dynamic mechanical properties: in torsion	
D5420	Test method for impact resistance of flat, rigid specimen by means of a striker impacted by a falling weight (Gardner impact)	
D3763	Test for high-speed puncture properties of plastics using load and displacement sensors	6603.7
D256	Test method for determining the Izod pendulum impact resistance of plastics	
D4812	Test method for cantilever beam impact strength of plastics	179
D695	Test method for compressive properties of rigid plastics	604
D2990	Test method for tensile, compressive and flexural creep and creep rupture of plastics	899
D648	Test method for deflection temperature of plastics under flexural load in the edgewise position	75
D1525	Test method for Vicat softening temperature of plastics	306
D3417	Test method for enthalpies of fusion and crystallization of polymers by differential scanning calorimetry (DSC)	
D3418	Test method for transition temperatures of polymers by differential scanning calorimetry (DSC)	
D696	Test method for coefficient of linear thermal Expansion of plastics between −30 °C and 30 °C with a vitrious diliatometer	3167
D2732	Test method for unrestrained linear thermal shrinkage of plastic film and sheeting	11501
D2838	Test method for shrink tension and orientation release stress of plastic film and thin sheeting	

ASTM	Title	ISO Equivalent
D1204	Test for linear dimensional changes of nonrigid thermoplastic sheeting or film at elevated temperatures	
D1238	Test method for melt flow rates of thermoplastics by extrusion plastometer	1138
D4440	Test method for plastics: dynamic mechanical properties: melt rheology	

Index